Civil War Torpedoes and the Global Development of Landmine Warfare

WAR AND SOCIETY

Series Editors: Michael B. Barrett and Kyle Sinisi

The study of military history has evolved greatly over the past fifty years, and the "War and Society" series captures these changes with the publication of books on all aspects of war. The series not only examines traditional military history with its attention to battles and leaders, but also explores the broader impact of war upon the military and society. Affecting culture, politics, economies, and state power, wars have transformed societies since the ancient world. With books that cut across all time periods and geographical areas, this series reveals the history of both the conduct of war and its societal consequences.

Gabriel Baker, *Spare No One: Mass Violence in Roman Warfare*

Marc Gallicchio, *The Scramble for Asia: U.S. Military Power in the Aftermath of the Pacific War*

Earl J. Hess, *Civil War Torpedoes and the Global Development of Landmine Warfare*

Brian D. Laslie, *Air Power's Lost Cause: The American Air Wars of Vietnam*

Geoffrey Megargee, *War of Annihilation: Combat and Genocide on the Eastern Front, 1941*

Lawrence Sondhaus, *German Submarine Warfare in World War I: The Onset of Total War at Sea*

Haruo Tohmatsu and H. P. Willmott, *A Gathering Darkness: The Coming of War to the Far East and the Pacific, 1921–1942*

Alan Warren, *Slaughter and Stalemate in 1917: British Offensives from Messines Ridge to Cambrai*

H. P. Willmott, *The War with Japan: The Period of Balance, May 1942–October 1943*

Thomas W. Zeiler, *Unconditional Defeat: Japan, America, and the End of World War II*

Civil War Torpedoes and the Global Development of Landmine Warfare

Earl J. Hess

ROWMAN & LITTLEFIELD
Lanham • Boulder • New York • London

Published by Rowman & Littlefield
An imprint of The Rowman & Littlefield Publishing Group, Inc.
4501 Forbes Boulevard, Suite 200, Lanham, Maryland 20706
www.rowman.com

86-90 Paul Street, London EC2A 4NE

Copyright © 2023 by Earl J. Hess

All rights reserved. No part of this book may be reproduced in any form or by any electronic or mechanical means, including information storage and retrieval systems, without written permission from the publisher, except by a reviewer who may quote passages in a review.

British Library Cataloguing in Publication Information Available

Library of Congress Cataloging-in-Publication Data

Names: Hess, Earl J., author.
Title: Civil War torpedoes and the global development of landmine warfare / Earl J. Hess.
Description: Lanham: Rowman & Littlefield, [2023] | Series: War and society | Includes bibliographical references and index.
Identifiers: LCCN 2022053850 (print) | LCCN 2022053851 (ebook) | ISBN 9781538174272 (cloth) | ISBN 9781538174289 (paperback) | ISBN 9781538174296 (epub)
Subjects: LCSH: Land mines—Confederate States of America. | Torpedoes—Confederate States of America. | Submarine mines—Confederate States of America. | Confederate States of America. Army—Weapons systems. | Land mines—History.
Classification: LCC UG490 .H483 2023 (print) | LCC UG490 (ebook) | DDC 355.8/25115—dc23/eng/20221108
LC record available at https://lccn.loc.gov/2022053850
LC ebook record available at https://lccn.loc.gov/2022053851

To Pratibha and Julie,
with love

Contents

Abbreviations	ix
Preface	xiii
Chapter 1: Before the War: "A Popular Aversion to Their Use"	1
Chapter 2: Lexington, Port Royal, Columbus: "All That Is Really Cowardly, Treacherous, Base and Brutal"	13
Chapter 3: Yorktown: "The Horrible Realities of War"	21
Chapter 4: Vicksburg, Port Hudson, Jackson: "Legitimate Weapons of Warfare"	37
Chapter 5: Fort Wagner, Fort Esperanza: "Scarcely Civilized Warfare"	51
Chapter 6: Sheridan's Raid, Petersburg: "These Shells Are Now Appreciated"	67
Chapter 7: March to the Sea, Pooler Station, Fort McAllister: "This Was Not War, but Murder, and It Made Me Very Angry"	81
Chapter 8: Fort Fisher, Sister's Ferry, Carolinas: "This Low and Mean Spirit of Warfare"	99
Chapter 9: Mobile: "A Thundering Report, a Flash, and the Groans of Wounded Men"	113
Chapter 10: Attacking Communications: "He Did Not Want the Matter to Become Public"	131
Chapter 11: Developments in the United States after the Civil War: "The Weapon the Father Prepared, May Turn Against the Son"	155

Chapter 12: Global Developments after the Civil War: "The
    Effective Stigmatization of the Weapon" 175

Conclusion 185

Appendix: A Medical Perspective on Landmine Injuries 193

Notes 197

Bibliography 223

Index 243

About the Author 251

Abbreviations

ALPL	Abraham Lincoln Presidential Library, Springfield, Illinois
BC	Bowdoin College, Special Collections and Archives, Brunswick, Maine
BHL-UM	University of Michigan, Bentley Historical Library, Ann Arbor
CALS	Cheshire Archives and Local Studies, Chester, England
CHM	Chicago History Museum, Chicago, Illinois
CinHM	Cincinnati History Museum, Cincinnati, Ohio
CWM	College of William and Mary, Manuscripts and Rare Books Department, Williamsburg, Virginia
DU	Duke University, Rubenstein Rare Book and Manuscript Library, Durham, North Carolina
EU	Emory University, Manuscript, Archives, and Rare Book Library, Atlanta, Georgia
EWU	Eastern Washington University, Archives and Special Collections, Cheney, Washington
FHS	Filson Historical Society, Louisville, Kentucky
HL	Huntington Library, San Marino, California
HU	Harvard University, Houghton Library, Cambridge, Massachusetts
LC	Library of Congress, Manuscript Division, Washington, DC
LMU	Lincoln Memorial University, Abraham Lincoln Library and Museum, Harrogate, Tennessee
LSU	Louisiana State University, Louisiana and Lower Mississippi Valley Collection, Special Collections, Baton Rouge, Louisiana
MaryHS	Maryland Historical Society, Baltimore, Maryland
MDAH	Mississippi Department of Archives and History, Jackson, Mississippi
MHM	Missouri History Museum, St. Louis, Missouri
MHS	Massachusetts Historical Society, Boston, Massachusetts

MML	Mariner's Museum Library, Newport News, Virginia
MU	Miami University, Walter Havighurst Special Collections, Oxford, Ohio
NARA	National Archives and Records Administration, Washington, DC
NC	Navarro College, Pearce Civil War Collection, Corsicana, Texas
NL	Newberry Library, Chicago, Illinois
NYPL	New York Public Library, Rare Books and Manuscripts, New York, New York
OHS	Ohio Historical Society, Columbus, Ohio
OR	*The War of the Rebellion: A Compilation of the Official Records of the Union and Confederate Armies.* 70 vols. in 128. Washington, DC: Government Printing Office, 1880–1901
ORN	*Official Records of the Union and Confederate Navies in the War of the Rebellion.* 30 vols. Washington, DC: Government Printing Office, 1894–1922
RNB	Richmond National Battlefield, Richmond, Virginia
SCHS	South Carolina Historical Society, Charleston, South Carolina
SIU	Southern Illinois University, Special Collections Research Center, Carbondale, Illinois
SOR	*Supplement to the Official Records of the Union and Confederate Armies.* 100 vols. Wilmington: Broadfoot Publishing, 1995–1999
SU	Stanford University, Special Collections and University Archives, Palo Alto, California
TSLA	Tennessee State Library and Archives, Nashville, Tennessee
TU	Tulane University, Special Collections, New Orleans, Louisiana
UGA	University of Georgia, Hargrett Rare Book and Manuscript Library, Athens, Georgia
UK	University of Kansas, Spencer Research Library, Lawrence, Kansas
UNC	University of North Carolina, Southern Historical Collection, Chapel Hill, North Carolina
UND	University of Notre Dame, Rare Books and Special Collections, South Bend, Indiana
UO	University of Oklahoma, Western History Collections, Norman, Oklahoma
US	University of the South, Sewanee, Tennessee
USAMHI	US Army Military History Institute, Carlisle, Pennsylvania

USC	University of South Carolina, South Caroliniana Library, Columbia, South Carolina
UVA	University of Virginia, Special Collections, Charlottesville, Virginia
VNMP	Vicksburg National Military Park, Vicksburg, Mississippi
WHS	Wisconsin Historical Society, Madison, Wisconsin
WLC-UM	University of Michigan, William L. Clements Library, Ann Arbor, Michigan
WRHS	Western Reserve Historical Society, Cleveland, Ohio

Preface

The Civil War took place at the start of a new phase in the global history of landmines. Prior to the nineteenth century, antipersonnel devices planted a few inches into the surface of the ground had been crude affairs called fougasses. They were little more than powder charges placed at the bottom of holes with stones above the powder to inflict injury. Set off by a lanyard pulled by an operator, they depended on exact timing to catch an attacking wave of troops at the right moment and were always placed immediately in front of a fortified position. Evidence of their existence is ample in the manuals written by military men when discussing the attack and defense of fortresses, but evidence that they were used in actual operations is very scanty. What evidence exists suggests they seldom were effective.

The era of the fougasse, which lasted from the sixteenth century to the early nineteenth century, slowly gave way to the modern era of landmine warfare as new technology and a handful of men in different countries began to have an impact. Technically, the device improved a great deal. Rather than a hole filled with powder and stones, it became a premanufactured object, either designed and made with the intention of being placed in the ground or improvised by taking an artillery shell and planting it as a landmine. In both cases it was a self-contained device that could be more quickly and easily placed. In terms of exploding it, the complicated mechanism of stretching a lanyard from the fougasse into the fort and waiting for the right moment to pull the cord was replaced by a victim-activated device.

These two basic changes—a prefabricated assembly for easy planting and a device that transformed the victim into his own executioner—revolutionized landmine warfare. They took place in the middle of the nineteenth century in two countries, one a relatively backward nation engaged in a short but intense war on its periphery and the other, one of the more technologically advanced nations engaged in a deadly internal conflict. The Russians deployed the first prefabricated, victim-activated landmine to protect their position during the siege of Sebastopol in the Crimean War of 1854–1855. The Confederates deployed their own version of this type of landmine at several locations during their war against the United States government in 1861–1865. In both wars the development of a new type of landmine depended on a small handful of men, who often acted against the inclinations of their comrades, but who gained the support of their military and civilian leaders. In both cases, landmine use created a technological precedent, but only in the Civil War did it also create a doctrinal precedent. The man responsible for generating Confederate landmine use wrote the first treatise in global history that detailed how and when to use the devices and how to plant them in effective and orderly fashion.

All of this may influence the student of military history to think that Sebastopol and the Civil War started the ball rolling toward the development of modern methods of landmine warfare. To a degree that is true, but equally important, it is a false assumption. While engineers were fascinated with the Russian landmines and detailed them in technical reports, the Civil War drew virtually no such interest from the many foreign military observers who came to the United States to witness the conflict. If they noted mines, they were far more likely to pay attention to underwater devices rather than landmines. Submarine devices were used much more widely and effectively in the Civil War and thus tended to grab more attention than land devices.

But that does not mean the Civil War lacked relevance in the global development of landmine warfare. Unlike Sebastopol, the Civil War led to theories about deployment and use of these devices that presciently identified the key components of a landmine doctrine. Those theories lay in an unpublished book authored by Gabriel J. Rains, the Confederate mastermind of landmine warfare, until Civil War historian Herbert M. Schiller edited it for publication in 2011. Its tenets would have been completely familiar with the military men of the 1930s and 1940s, who created the full doctrine of landmine warfare to go along with the first modern landmine developed in Germany, the Soviet Union, and other participants of World War II.

It is not easy to understand why what had been introduced in 1854 in Russia and 1861 in the United States took so long to be recognized, copied, and further developed in those and other countries. Not even the long, static operations along the western front of World War I led to significant use of landmines. But the history of these devices indicates a long period of disinterest followed, in the era of World War II, by an explosive reversal of that attitude. Landmines were used by the millions in that war and continued to expand around the world during the Cold War to become one of the most persistent and tragic problems faced by the global community.

The Civil War played a role in this long history that deserves to be brought into focus. When technicians and military planners worked out methods for mass-producing modern devices by the thousands and developed a workable doctrine for when and where to use them and how to organize the planting, they essentially duplicated, without realizing it, the basics of Rains's thinking. The story of the Civil War and its role in the global development of landmine warfare is not a story of cause and effect. It is a story of concurrent development, of military men working out a new procedure on their own based on trial and error, without reference to what had been or had not been tried before. The history of war technology more often plays out that way than in a simple, straightforward system of transference of information from one developer to another.

For good or for ill, the Confederacy was a significant player in the early stages of landmine development, and the Federal reaction to the employment of these devices by their enemy was thoroughly negative. The first deadly use of landmines took place on the rainy morning of May 4, 1862, as thousands of Federal troops moved forward through the Confederate line of fortifications based in Yorktown that had just been evacuated the night before. Some of them stepped on explosive devices that went off with devastating effect, killing and horribly mangling several men. Northern soldiers and civilians alike were so morally outraged by their use that they never forgot the torpedo incidents at Yorktown. For them, it was yet another example of barbarism in Southern culture, another indication that their enemy did not obey the laws of war, and another reason to fight them to the last.

This study details the history of landmine development and use in the Civil War by focusing on four themes. Three of them are morality, tactics, and technology, and placing Civil War landmines within the context of global history is the fourth purpose of this study. All four themes were connected and are essential for a full understanding of this controversial weapon of modern warfare. This book deals only with landmines, or what the Civil War generation called sub-terra shells. The underwater mines, antiship rather than antipersonnel devices, have a history that is very different from that of landmines. In fact, submarine mines deserve a separate book-length study that also places the Civil War within the larger story of global military history.

From Port Royal, South Carolina, in November 1861, to the campaign for Mobile in March and April 1865, a total of fifteen instances of landmine use occurred during the Civil War, all of them by the Confederates. In addition, there was a terrible accident at Batchelder's Creek, North Carolina, on May 26, 1864, in which some Federal underwater torpedoes exploded and killed or wounded 55 people. A Confederate act of sabotage at the Union supply base of City Point, Virginia, killed and wounded at least 136 men on August 9, 1864. If one adds the casualties in these two incidents to the casualties incurred by the fifteen instances of landmine use, the total of killed and wounded amounts to at least 319 losses. That figure is based on evidence that is reasonably dependable, as opposed to rumors or unconfirmed reports, and must be taken as the minimum number of torpedo casualties during the Civil War. Among that number were a handful of Confederate soldiers and civilians, the latter of whom included some women and children.

Prior work on the history of landmines in the Civil War began with Milton F. Perry's *Infernal Machines: The Story of Confederate Submarine and Mine Warfare*. Perry's book has served as the standard history of both landmines and underwater mines during the Civil War for many decades. It deals mostly with the technology of this weapon with slight discussion of tactics and morality and has, for many years, been out of print. While Perry covered both landmines and underwater mines, he paid more attention to the latter.

Mike Croll's *The History of Landmines* is not specifically a Civil War study but a general history of landmines in global history. Croll, an officer in the British army who later became a self-styled "humanitarian deminer," paid some attention to the Civil War but mostly devoted his study to the twentieth century. While working to clear minefields in Cambodia during the early 1990s, Croll became interested in the history of this weapon. "I approach the subject from an informed but opinionated position," he has written. "Today it is impossible to cover this subject without reference to the humanitarian perspective." Croll calls the landmine "one of the most insidious weapons ever developed."[1]

Norman Youngblood's *The Development of Mine Warfare: A Most Murderous and Barbarous Conduct* is another general history of the subject with brief attention paid to the Civil War. Youngblood also pays a bit more attention to the moral argument about landmines than does Croll.

Confederate Torpedoes: Two Illustrated 19th Century Works with New Appendices and Photographs by Herbert M. Schiller (the editor) brings to publication Gabriel J. Rains's "Torpedo Book," written about 1874, which explains the general's conception of landmine warfare. Schiller provided very helpful editorial commentary on that document. He also edited a long report written by Union engineer Peter S. Michie about Confederate torpedoes. Completed by October 1865, this report also had never been published before Schiller included it in his book. The editor also graced both reports with extensive illustrations.

Michael P. Kochan and John C. Wideman's *Civil War Torpedoes: A History of Improvised Explosive Devices in the War Between the States* also deals with the subject. Like Perry, their primary focus is on technology but with deeper research and a more assured discussion because they utilized a wider array of sources than Perry. But Kochan and Wideman did not delve much into the moral arguments or the tactical deployment of these weapons.

The latest book that deals with Civil War torpedoes is Kenneth R. Rutherford's *America's Buried History: Landmines in the Civil War*. Written by a humanitarian relief worker who lost both his legs to a landmine in

Somalia in 1993, it is a brief introduction to the subject. Rutherford based his book on minimal research, especially in archival sources, misses the Pooler Station incident entirely, and too readily relies on Confederate sources in evaluating the tactical effectiveness of various landmine incidents. Also, there is no larger context before 1861 or after 1865 in his study, no linking Civil War developments with the global development of landmine warfare.

For other studies that deal with land and underwater mines in the Civil War, see Francis A. Lord's "Both Sides Used Torpedoes Widely"; Louis S. Schafer's *Confederate Underwater Warfare: An Illustrated History*; Dean Snyder's "Torpedoes for the Confederacy" in *Civil War Times Illustrated*; and Timothy S. Wolters's "Electric Torpedoes in the Confederacy: Reconciling Conflicting Histories" in the *Journal of Military History*.

While I learned a good deal from the work of all previous historians of the landmine, I have to disagree with them on a relevant issue of classification. There is a tendency in the historiography to conflate landmine warfare with any other type of conflict that involved explosives planted under the surface of the earth. But there are important differences in the technology and the tactical use of various underground devices. The biggest problem lies in the use of the word *mine*, for its meaning changed over time to create confusion in the minds of some historians.

The best example of this confusion lies in the practice of military mining compared to landmine use. Military mining, as ancient as organized warfare itself, was an aspect of siege operations that involved digging a tunnel under an enemy fortification and collapsing it or, with the coming of the gunpowder age, blowing it up. This had nothing to do with planting small devices a few inches under the ground to kill or injure enemy troops, yet numerous historians have discussed both operations as if they were the same thing.

Likewise, previous historians of the landmine have often connected this device to the age-old practice of placing obstructions on the surface of the earth to defend a fortification. These obstructions include caltrops (spiked devices to injure horses' hooves) or pits with sharpened stakes. While the purpose of these devices was similar in some ways to that of landmines, their technical aspects were so different as to make comparison unimportant.[2]

For the sake of clarity, when I use the term *landmine warfare*, I am referring only to the use of an explosive device designed to harm people and planted a few inches below the surface of the earth. After the Civil War landmines began to be planted for the purpose of harming material (such as tanks) as well.

Also, for the sake of clarity, a word about terminology is appropriate. The Civil War generation commonly used the term *torpedo* to denote any explosive device designed to injure people or material. They also used variations on that term such as "torpedo mines."[3]

The term *torpedo*, however, has a very different connotation today because of developments that began to take place immediately after the Civil War. By the later 1860s, a British inventor developed a device that carried a powder charge through the water toward its target, using compressed air as the motive power. This device came to be exclusively called a torpedo by the turn of the twentieth century, while the term *landmine* came to be exclusively used for what Civil War contemporaries had previously called a torpedo.[4]

The reason that members of the Civil War generation called explosive devices torpedoes was because of the shock and danger involved in the connotation of that term. The word derived from "the electric ray or skate," according to Lt. Commander John S. Barnes, who wrote one of the early books on torpedo warfare right after the Civil War. The sea creature was "said 'to kill its prey as by lightning.'"[5]

According to the *Oxford English Dictionary*, the torpedo is defined as a "flat fish of the genus *Torpedo* or family *Torpedinidae*, having an almost circular body with tapering tail, and characterized by the faculty of emitting electric discharges." It also was known as an "electric ray, cramp fish, cramp ray, and numb fish." The earliest reference to it, dated about 1520, reads, "Torpido is a fisshe, but who-so handelth hym shalbe lame & defe of lymmes that he shall fele no thyng." The word *torpedo* was first associated with underwater mines in 1776 and with landmines in a book called *Beckford's Valthek*, published in 1786. "I will spring mines of serpents and torpedos from beneath them, and we shall soon see the stand they will make against such an explosion."[6]

In the United States, the term *torpedo* was first used in the obituary of English author John Cleland that appeared in the *Gazette of the United States* in 1789. "As a writer, he showed himself best in novels, song-writing, and the lighter species of authorship; but when he touched politicks he touched it like a torpedo, he was cold, benumbing, and soporifick."[7]

George B. Prescott wrote of the fish that lent its name to Civil War landmines in his *History, Theory, and Practice of the Electric Telegraph*. It was published in Boston in 1860, only one year before the outbreak of the war. Prescott discussed several fishes reported to be capable of producing electricity. At Cape Cod, residents knew of the one called torpedo, calling it cramp fish. "The magnitude and number of nerves in the torpedo are very far greater than those supplied to any other animal," Prescott asserted. The "power of transmitting those powerful shocks of electricity is controlled and regulated by the will of the animal." Cape Cod residents insisted that merely touching the fish numbed their hands and arms. The shock for some was so great "as to prostrate a man instantly."[8]

The terminology associated with the word *torpedo* was clearly negative, denoting danger and bodily harm. When several torpedo crimes were publicized in Northern newspapers during the 1850s, spelling out in chilling detail how one person tried to kill another by sending them a bomb concealed in an ordinary package, another layer of horror was added to public perception of antipersonnel devices. Civil War soldiers and civilians alike tended to apply various terms to landmines, which denoted their moral revulsion against the weapon. They called them "infernal machines," "instruments of murder," and the "devil's 'paving stones.'"[9]

Torpedo use became so widespread and public awareness of it increased to such an extent that the word came to embrace many things for the Civil War generation. "The term has however, become so general in use as to include nearly all explosive apparatus, for war purposes, on land or sea, excepting that relating to ordinary fire arms, artillery and mines," concluded Capt. W. R. King. This Federal engineer officer authored the first book-length study of torpedo warfare to be published in America, and he wrote it in the immediate aftermath of the Civil War. As we have seen, the word *torpedo* eventually was streamlined of all these meanings and came only to be applied in military terms to a self-propelled device that traveled just below the surface of the water toward its target long after the Civil War.[10]

But later, during the twentieth century, the term *torpedo* took on added connotations based on its application to the sleek, silent, and piercing effect of a naval weapon that sliced into ships when the enemy least expected it. A professional gunman or bomber was called a torpedo. The term also was used less dangerously to denote a sleek car design or a very pointed beard, and a strong and dangerous mix of liquor was sometimes referred to as torpedo juice.[11]

In virtually every way it was used, the term *torpedo* had negative and dangerous connotations. The Civil War generation inherited that trend from previous generations, perpetuated it widely in their conflict, and bequeathed it to future generations of the twentieth century. The shock of its use at Yorktown and other locations sealed the anger and revulsion in the minds of Northerners, and those emotions never dimmed until the Civil War generation was no more.

In the preparation of this book, I wish to thank Herbert M. Schiller for sharing with me the results of his research and his thoughts about the subject and, especially, photographs of landmine specimens currently housed in various museums.

Mostly, I want to thank my wife, Pratibha, for her love and support.

Chapter 1

Before the War

"A Popular Aversion to Their Use"

Long before the American Civil War, developments in many parts of the world led to a comparatively primitive device that presaged the landmine. Appearing as early as the ninth century in China, gunpowder eventually gave rise to the ancestor of the modern landmine, a device known as the fougasse. The Chinese created a version of it by 1277, which was detonated either by pulling some sort of friction device or by pressure placed on some sort of match to touch it off. The fougasse became more common one hundred years later in Chinese warfare with trip-wire detonation. The Chinese also developed a water mine that could float along with the current and which was detonated with a time-delay fuse.[1]

The first recorded use of a fougasse in European history dates to 1573, when Samuel Zimmerman created one during the siege of Augsburg. Zimmerman placed it on the glacis of a fortification and arranged a flintlock device to detonate it. He also arranged a trip wire to activate the flintlock, creating a "target-activated" antipersonnel weapon.[2]

Fougasses were used at least sporadically during the eighteenth century—for example, at the siege of Schweidnitz in 1762 during the Seven Years' War. Although not common, they found their way into technical manuals during the latter decades of the century. The manuscript of an unidentified Prussian officer, translated into English by Capt. J. C. Pleydell of the Fifth Regiment Foot Guards, gave ample attention to the device. Published six years after the siege of Schweidnitz, the Prussian manuscript recommended that fougasses be planted to screen weak portions of a fortification, such as salients and portions not covered by cross fire. They were best placed ten to fourteen feet from the outer ditch in pits three feet square and six to eight feet deep. A hole was dug at the bottom to create a chamber for the gunpowder.[3]

The key to the fougasse was a proper detonation device, and the Prussian officer wrote in great detail about that topic. He described a pipe made of "coarse cloth" that was two inches in diameter with half a pound of gunpowder per linear foot, which was laid in a wooden trough. It stretched from the powder chamber at the bottom of the pit to the fortification. Lighting the end of the pipe to set off the fougasse, the Prussian officer had great faith in the reliability of the device. He estimated that four men could construct a fougasse in six to seven hours.[4]

While the fougasse appeared in technical manuals, they were rarely deployed and even more rarely successful. The primary problems were the time and expertise needed to make them and the unreliable nature of the detonating device. Even if the latter worked technically, there was the question of timing. A defender had to blow up the fougasse exactly when approaching troops were near enough for it to have a physical effect on them. A victim-activated device would solve that problem, but no one other than Zimmerman seems to have deployed such a device.

J. G. Tielke, a Saxon captain of artillery, admitted in a book about field engineering published in 1789 that fougasses were "much more harmless than is generally imagined." But he asserted that they were worth the trouble because of the emotional effect they could have on the enemy. If the opponent knew where they were planted, he would not likely approach that spot "or proceed with the greatest irregularity and hesitation."[5]

In fact, there is evidence for the increasing use of fougasses by the late eighteenth and early nineteenth centuries. A permanent system of fougasses was created on the island of Malta by about 1770 to defend possible landing sites. Although there is no evidence that a fougasse was created in North America before the American Revolution, a French engineer set up one at Fort Mercer in New Jersey in October 1777. They were used reportedly with significant effect at the sieges of Ciudad Rodrigo, Badajoz, and Santander during the Peninsular War of the Napoleonic era.[6]

The only time that American troops had to contend with a fougasse occurred during the Mexican War. As part of the defenses of Chapultepec, the Mexicans dug at least six fougasses and charged three of them before the Americans attacked. Placed at the west wall of the enclosed compound protecting this military college, the fougasses were positioned outside a defensive ditch on top of a hill. At the base of the slope was a grove of cypress trees.[7]

The Americans attacked on the morning of September 13, 1847, and easily cleared the grove of Mexican troops. Then the Ninth US Infantry advanced up the hill, but the men carrying scaling ladders had not kept pace with the forward advance. This compelled the troops to fall back and take cover, firing at the top of the wall for at least fifteen minutes. Engineer Lt. P. G. T. Beauregard noticed "a certain number of *mounds of earth and stone*," on the ground, which he assumed were graves. Later he realized, "to [his] *horror*," that they were fougasses. How many other Americans realized the same is not known, but after a quarter-hour delay, reinforcements arrived in the form of the Fifth, Sixth, and Eighth US Infantry, along with the scaling ladders. Now the Americans pushed up and over the masonry wall and into the enclosure, capturing the Mexican position by 9:30 a.m.[8]

Mexican engineer Lt. Alemán had been assigned the task of touching off the fougasses but failed to do so. In his trial after the battle, Alemán explained that he tried to perform his duty but found his way to the firing station blocked by retreating Mexican troops. Americans had already gained possession of the station before he could reach it. Alemán also asserted that at least some of the canvas pipes containing the powder trains had been cut by the attackers by that time as well.[9]

The experience at Chapultepec neatly illustrated the problem with fougasses. Not only was the detonation method unreliable, but there was an issue concerning the exposure of gunpowder for a length of time in a hole in the ground. The powder train also would have been exposed to moisture although encased in a wooden trough. Even if the Chapultepec fougasses had all gone off, it is doubtful they would have played a large role in reversing what was already a desperate situation for the Mexican defenders. As historian Mike Croll has concluded more generally of its history, the fougasse "was rarely decisive and frequently unreliable."[10]

Nevertheless, military writers of the Civil War era devoted attention to the fougasse. Henry W. Halleck described the device in his *Elements of Military Art and Science*, published originally in 1846, a year before the incident at Chapultepec. He noted that it would be possible to detonate a fougasse by use of wires connected to a galvanic battery. This had not yet been done by 1846, but it held a great deal of promise. If workable, an electrical detonation system would solve all problems connected with the timing of the explosion. Col. Henry Lee Scott's *Military Dictionary* of 1861 largely duplicated Halleck's description of the fougasse. Scott admitted that the main problem with this elaborate device was "to explode them at the instant when the enemy is passing over," but he did not advocate the use of electricity. Instead, Scott suggested placing some obstacle to troop movement over or near the fougasse to stall the enemy where they could be most affected by the blast.[11]

RAINS AT FORT KING

Very few people knew it at the time, but an American army officer had already experimented with a reliable system of detonation a few years before the appearance of Halleck's book. From the standpoint of the defender, a victim-activated detonation system was the ultimate in reliability. Capt. Gabriel J. Rains of the Seventh US Infantry deployed such a device during the

Second Seminole War in Florida in the spring of 1840. Word of this incident was not widely known, not by intent but by default. There simply was no impetus for spreading information about it, and the incident lay in obscurity for many years.

Born at New Bern, North Carolina, in 1803, Rains graduated thirteenth in his class of thirty-eight cadets at the US Military Academy at West Point in 1827. He served as a lieutenant of infantry in the Indian Territory, participating in the removal of the Choctaw from Mississippi. In February 1835, his colleague Lt. Jefferson Davis was tried by a general court martial at Fort Gibson, Indian Territory. He was charged with absence at roll call and disrespectful language and conduct when confronted with his dereliction. Rains served on the court martial, which found him not guilty. This may have helped Rains nearly thirty years later when he needed the support of Davis, who then was president of the Confederate government. Rains married the granddaughter of Tennessee pioneer John Sevier in October 1835 and soon after was transferred to Florida where he participated in the Second Seminole War.[12]

As commander of Fort King in northern Florida, Captain Rains became angry at the skulking way of war practiced by the Seminole. On March 24, 1840, he lost two good men, who "were waylaid and assassin-like, shot down in sight of this Post." Rains decided to exact revenge. A hammock located two miles south of the post was a likely venue for a booby trap. Rains put a howitzer shell in a box with some loose gunpowder and pieces of old iron, buried it at the hammock, and placed a shirt belonging to one of the slain men on top. He then fixed it so that removing the shirt would trigger the explosive.[13]

Several evenings later, the garrison heard an explosion at the hammock, but on investigating, Rains found no one. He planted a second booby trap, which exploded on the evening of April 27, 1840; again he found no one in the vicinity. The next morning Rains led sixteen men, all he could spare of the fort's garrison, to scour the hammock which was 100 yards long and 140 yards wide. Rains deployed his small group as skirmishers and moved through the hammock. They encountered a large force of Seminole, up to one hundred warriors, who tried to cut off their retreat to the fort. Rains was shot through the lung in the hot skirmish that followed and began to feel faint from loss of blood. As three men carried him back to the fort, the others covered him and continued to press forward until they reached the safety of Fort King. Rains eventually recovered from his serious wound and continued on duty.[14]

Rains used existing artillery ordnance for his two booby traps at Fort King. A howitzer shell with a friction fuse designed to explode the shell when striking a solid object in flight was all he needed. He did not explain exactly how he arranged it so that pulling on the clothing would detonate the shell, but it was done with available material at the fort. In modern parlance it would be called a disturbance fuse.[15]

There is no record of using friction fuses to set off a torpedo before Rains's innovation in 1840. There also is no indication that Rains gave any further thought to what he had done until the Civil War inspired him to build on his experiment during the Second Seminole War. He could not have known it at the time, but using a contact-activated device to explode torpedoes was the wave of the future in landmine warfare.

ELECTRICAL DETONATION

Detonating underground explosives by means of electricity also was a new idea, but developing a reliable method of doing so proved to be a long and difficult problem. Harnessing electrical power for anything was in its infancy during the mid-nineteenth century. Initial experiments by Prof. Luigi Galvani of the University of Bologna produced a scientific dialogue on the nature of electricity during the 1790s. At the same time, Prof. Alessandro Volta at the University of Pavia created a device to produce electric currents, publishing his results in 1799. The voltaic pile consisted of alternating copper and zinc discs arranged around a central stick, each disk separated by cardboard soaked in salt water to act as electrolytes. It was the first electrical battery designed to produce continuous current. This pile provided the basic concept of producing electricity for the next seventy years until the invention of the dynamo a decade after the Civil War.[16]

Often called galvanic batteries or electrochemical batteries, devices growing out of Volta's concept gave rise to two types of cells. Primary cells produced electricity and secondary cells stored it. The most famous example of the latter was the Leyden jar. Most applications of electricity to real-life tasks required a primary cell to create power, but secondary cells could be used to store and expend it. William Robert Grove of England improved on the basic electrochemical design in 1839, one of several to appear during the pre–Civil War years. Grove's became the most commonly used generator of electricity for telegraph work in the United States.[17]

An alternative to electrochemical batteries was the electromagnetic method of producing electrical power. Rather than relying on the reaction of various chemicals, this device relied on manipulating magnets to produce energy. It was initially developed by Prof. Hans Christian Øersted at the University of Copenhagen by 1820. The device was improved by several people, including Michael Faraday in England and Joseph Henry in the United States.[18]

At roughly the same time as the development of electrical devices, a handful of Americans had been working on the theory of exploding torpedoes as part of the nation's defense plans, but in the absence of reliable detonation methods, little had come of their work. Steamboat inventor Robert Fulton proposed a system of deploying underwater mines to protect American port cities. Publishing his proposal as *Torpedo War and Submarine Explosions* in 1810, Fulton convinced Congress to fund experiments, which failed. Weapons producer Samuel Colt proposed his own system of underwater mines detonated by means of a galvanic battery in the early 1840s. His experiments were successful, yet the army, navy, and Congress failed to fund or adopt his work.[19]

The Europeans were ahead of the Americans in applying electricity to torpedoes. Baron Pavel Schilling von Canstadt successfully touched off underwater mines by using electrochemical devices as early as 1812 in Russia. During the 1830s both electrochemical and electromagnetic devices were used to explode underwater charges designed to break up the wreck of the naval vessel *Royal George*, which had sunk near Portsmouth in 1782. The explosions facilitated salvage and clearance of the hulk. Experiments conducted by the British during the late 1850s concluded that electromagnetic current was superior to electrochemical energy to explode gunpowder. It could detonate up to a dozen charges at a time. Charles Wheatstone of England developed the best electromagnetic machine by adding more horseshoe magnets and more helices, turning the armature rather than the helices as in previous devices. By hand cranking the device, Wheatstone could provide reliable current to explode mines and submarine torpedoes, igniting from two to twenty-five charges at the same time.[20]

SEBASTOPOL

The first large-scale use of a modern landmine as opposed to a fougasse occurred in the Crimean War of 1854–1855. The Russians deployed several hundred such devices as part of their defensive system protecting the port city of Sebastopol, which was ringed with earthworks on the land side. The mines were detonated by a unique system of contact-activated fuses based on the mixing of chemicals, although it was not an electrochemical device for generating electricity.

The man responsible for these torpedoes was Immanuel Nobel, the progenitor of a Swedish family that rose to prominence in European industry and technology. An architect, inventor, and industrialist, Immanuel moved his family to St. Petersburg in 1838, where Tsar Nicholas I became interested in his underwater mine. Four of his sons followed in his footsteps, including Alfred Nobel, who later developed dynamite. Immanuel Nobel did not receive widespread recognition for his Sebastopol landmines because, for some reason, another émigré was credited with them. German-born Moritz Hermann Jacobi, a chemist, also had found favor with the Russian government. The most prominent member of the Russian Committee on Underwater Mines, most observers assumed Jacobi was the developer of these torpedoes. Actually, the committee called on Immanuel Nobel for his advice, which led to Nobel's development of the landmines. There is no evidence that Jacobi ever worked with explosive devices of any kind, his interest leading him to experiment with electromagnetic energy and developing a small, battery-powered boat.[21]

Both Nobel's underwater and sub-terra mines operated with the same detonation device, a system that mixed chemicals on the spot to create an explosion. In both cases, the powder charge, contained in a box, was positioned just below the surface of the water or the land with a metal tube sticking up. Inside the metal tube was a glass tube that broke when a vessel hit or a foot stepped on the metal tube. The intermixing of sulfuric acid, potassium, and sugar produced the desired explosion. Nobel's landmines contained anywhere from eight to twenty-five pounds of powder.[22]

The operations around Sebastopol gave ample opportunity for the deployment of Nobel's device. French and British forces planted themselves along the land side of the port city but were unable to cut it off from the outside world. They focused on heavy artillery bombardments, digging-siege approaches, and conducting periodic assaults to capture the place. A major attack was planned in which French troops would assault some outer works on the allied right, and British forces would capture other Russian outer works at the Quarries. After a severe preparatory bombardment, the allied move took place on the afternoon of June 7, 1855.[23]

British troops found that their opponent had laid hundreds of Nobel landmines barring their way to the Quarries. With luck, they managed to move across no man's land without detonating too many of them to have an effect on checking the assault. The British captured the outer works and consolidated their new position. While hurting only a handful of troops during the attack, the mines exacted a large toll after the assault had ended. Maj. Reynell Pack of the Seventh Fusiliers reported that Captain Armstrong of the Forty-ninth Regiment had been wounded in the assault. He was carried to the rear on a stretcher by four men, one of whom stepped on a torpedo. All members of the group were "more or less severely injured," but none of them immediately died as a result of the explosion.[24]

But many others were not so fortunate. Troops stumbled on these torpedoes for several days after the attack. English sappers dug up at least one hundred of them from June 7 to June 9 but were unable to find and take away all of those that lay between their old position and the new line. Lt. Hugh Robert Hibbert of the Seventh Royal Fusiliers realized that the tall grass which had grown up between the lines that spring served to hide many of the torpedoes, and anyone walking through the grass was in danger of stepping on a tube, which Hibbert reported as two or three feet in length. It was wise for everyone to refrain from crossing the open ground, but inevitably, someone would do so "on private speculation to see what he could get." The result was periodic explosions heard most often in the night. "We have had many men blown up by these things," Hibbert told his mother.[25]

Pack confirmed Hibbert's report. He described a particularly gruesome incident on the night of June 11, when a work detail of three hundred men led by Major Herbert of the Twenty-third Regiment crossed the open ground. The detail formed a line two hundred yards long and eight feet wide, which limited its exposure to the landmines, but one man near the rear of the column stepped on a device. Three men were killed, and "many more were wounded." Herbert narrowly escaped, "being covered with blood, and the debris of the killed. The loss itself was nothing compared to the moral depression it raised amongst the division, and those who witnessed the sight." Pack believed the Russians understood that their primary aim was to demoralize the enemy rather than destroy them physically. He emphasized "the dread engendered amongst the men by these diabolical, hidden, machines." Hibbert imagined that the Russians could hear these explosions in the night and probably laughed at their success in inflicting a sense of horror on their enemy.[26]

Many interested parties conducted an examination of these Nobel landmines after the fall of Sebastopol to the Allies in September 1855. Col. Richard Delafield, the leader of a US Army commission to observe operations in the Crimea, looked at one that had been dug up, defused, and was stored in the English engineer depot. A box of gunpowder eight inches in dimension was placed inside a larger box, with two inches of pitch between the boxes to serve as waterproofing. The top of the outer box was placed eight inches below the surface. The example Delafield so minutely described only had a glass tube. A board six inches wide and twelve inches long had been placed on the ground over this box so that anyone stepping on the board would depress it enough to break the glass tube just below. Delafield also reported that other mines had been uncovered, which contained the glass tube within a tin tube. He measured one such example and found that the tin tube was $19\frac{7}{8}$ inches long and $\frac{3}{8}$ of an inch in diameter. The glass tube within it was 18.25 inches long and 3.32 inches in diameter. These dimensions insured maximum effect in breaking the glass tube if the tin tube were mashed almost anywhere along its length.[27]

The Russians planted Nobel landmines at more places than just the Quarries sector. French troops had to contend with them at Mamelon, Malakoff, and Mast Bastion. "Very many were placed throughout the defenses," asserted Delafield. He reported at least one instance of casualties among French sappers who tried to dig up these devices after the effort to capture positions had ended.[28]

Nobel's detonation system was unique. Delafield refrained from passing judgment on it, but he was interested in European efforts to employ electricity for detonating explosives. He found in Europe a paper written by Spanish engineer Col. Gregorio Verdú y Verdú on the use of electricity in military mining, or the digging of approach tunnels during sieges. An American artillery officer translated it for Delafield, who felt that the results, particularly over long distances, were still uncertain. He also wondered if any currently available electrical device could set off multiple charges. Delafield had a basic faith in electrical detonation systems. He knew that the French and British had used them extensively to destroy dock facilities at Sebastopol after the fall of the city. They blew up dry docks, warehouses, barracks, crenellated walls, and masonry forts from October 1855 until February 1856. The French and British used electrochemical devices to generate the charges for these explosions. While the British engineers were not fully satisfied with those devices, Delafield found that Austrian engineers were enthusiastic about them. An Austrian engineer, Lt. Col. Scholl, had developed an effective fuse (also called a deflagrator or a squib) to connect the electrical wiring to the powder charge and gave one to Delafield for study.[29]

Nobel contributed to a historic first at Sebastopol, the deployment of a large number of modern landmines in an active campaign. Russian engineers placed them, and French and British troops had to deal with them. While technically successful in reliably going off on contact with enemy troops, it cannot be said that the Nobel landmines were effective in a tactical sense. They failed to hinder, much less stop, the British assault on the works near the Quarries. The French assaulting columns that had to deal with them also were not deterred or hindered much in their progress. They had more effect on the enemy after the attacks than during them, randomly killing a number of French and British troops who were unaware of the danger inherent in wandering across what had once been no man's land between the lines. The torpedoes certainly had an emotional effect on the French and British soldiers, but those men largely absorbed the threat into their overall consciousness without becoming overly demoralized. In short, they now had another danger to worry about, along with Russian bullets, artillery fire, and the rampant disease that took far more lives among the Allies than combat-related causes in the Crimea.

The Russian deployment of underwater mines was far more successful than the landmine experiment. Nobel mines were deployed by the hundreds around island fortresses at Sveaborg and Kronstadt to protect the naval approaches to St. Petersburg. A British expedition along this route turned back for fear of running onto these mines in June 1854. When a joint French-British naval expedition attempted the same approach in the summer of 1855, the Allies engaged in the first mine-sweeping operation in naval history. They discovered forty-four underwater mines designed to explode by observers on shore who used electrical detonation systems. An additional 950 underwater torpedoes were found that were designed to be touched off by contact with vessels. British personnel managed to take up about fifty mines but, in the process, suffered damage to four ships. The mines were not deployed so as to form a barrier to movement but were scattered about in potential anchorages and, thus, were more of a harassment feature than a shield.[30]

With the deployment of Nobel underwater mines at several places in the Black Sea as well as along the approaches to St. Petersburg and the deployment of sub-terra mines along key sectors of the Sebastopol land defenses, one can see that Tsar Nicholas I envisioned torpedoes as important new weapons. One of the most technically backward nations of Europe was leading the way in this new military technology only because of the Swede Immanuel Nobel, who sold his inventions to the Romanov dynasty. Nicholas was so impressed that he continued to purchase Nobel mines after the Crimean War. The British government also was impressed and hired engineer William Armstrong to develop underwater mines for its navy after 1855.[31]

The United States was behind the curve in adopting the new technology. Delafield wrote of sub-terra torpedoes as if they were a new idea to him. Even though he minutely described the Nobel landmines at Sebastopol, he did not recommend them for use by the US Army. Delafield only partially endorsed the use of electrical devices to fire underground charges in siege warfare, indicating more testing was necessary to better gauge its effectiveness at exploding long-distance and multiple charges.

This is as far as the American military establishment had gone in its awareness of landmines by the time of the Civil War. Delafield reported that cadets at the US Military Academy were practiced in the use of electrical detonation for underwater charges and on land to set off large, underground charges in siege warfare. But no serious instruction or practice in landmines existed at West Point. Cadets were briefly introduced to the concept and fired off a practice round or two, but nothing else was done by the army to prepare for the most modern form of detonating charges (by electricity), and nothing was done to develop an American landmine.[32]

MORAL REVULSION

Perhaps the true reason for this lack of movement toward the new weapon was a considerable degree of moral revulsion against it. In 1878, long after the Civil War, Admiral David D. Porter wrote the article "Torpedo Warfare" for the *North American Review*. He identified the Crimean War as "the first instance we have where the humanitarian principle was wholly disregarded, and the torpedo openly made use of by a nation for offensive and defensive purposes." Porter reflected a general American attitude that torpedoes were morally reprehensible to a civilized people. It also needs to be noted that Porter was writing mostly, if not solely, about underwater mines rather than landmines. Even though underwater torpedoes were antiship rather than antipersonnel weapons, they still raised the ire of concerned Americans, who tended to view them as barbaric devices.[33]

Americans were ready to view the use of torpedoes as a criminal act because a couple of prominent cases of attempted murder using these engines of destruction became top news. The *New York Herald* reported on June 3, 1850, that an explosive device inside a box was sent to the home of Thomas Warner of that city. It exploded inside the house but failed to kill anyone. Investigators pinpointed Warner as the instigator, assuming he meant to murder his wife, and the newspaper labeled the incident a "torpedo crime." A shadowy figure called Mr. Thompson was identified by some informers as the maker of the bomb, but Warner vehemently denied any connection with the incident.[34]

In October 1852, a second incident targeted James Gordon Bennett, the editor of the *New York Herald*. A box allegedly containing samples of ore from Cuba and addressed to Bennett showed up at the Herald office. An alert employee decided to cut into it with a knife, and gunpowder began to spill out. He then soaked the box in water for several hours to neutralize the device. It was arranged with pellets of paper and friction matches so that when the top was disturbed, it would go off. Newspaper reports labeled it "an infernal machine," and yet another torpedo crime made headline news across the United States. It is possible the same mysterious Mr. Thompson was behind this as well, although no one seems to have identified him in connection with it.[35]

Negative connotations concerning the use of torpedoes prevailed in pre–Civil War America. It is true that the terms "firecrackers" and "torpedoes" were used to refer to noise-making devices exploded during Fourth of July celebrations every year, but those were considered harmless. Torpedo crimes in the civilian world and attempts to blow up unsuspecting soldiers after a battle had ended seemed beyond the pale of civilization. When naval officer W. R. King wrote the first government study of torpedoes in the aftermath of the Civil War, he speculated on why Americans had done so little to study torpedo warfare before the firing on Fort Sumter. King blamed it on "a popular aversion to their use, excepting in cases of absolute necessity." Until the Civil War broke out, he implied, there had been no such cases of "absolute necessity."[36]

Chapter 2

Lexington, Port Royal, Columbus

"All That Is Really Cowardly, Treacherous, Base and Brutal"

During the first year of the Civil War, only furtive efforts were made to use improvised landmines. In fact, it is not even certain that they appeared at the siege of Lexington, Missouri, in September 1861, and there is only limited evidence of their appearance in the defense of Port Royal, South Carolina, in November of that year. But there is overwhelming evidence that the Confederates deployed a system of land torpedoes as well as underwater mines at Columbus, Kentucky, during the late fall and winter of 1861–1862. Yet at none of these places were the mines actually touched off. While the Columbus torpedoes gained quite a bit of publicity, the fact that they hurt no one greatly limited public awareness of their existence. Not until a large deployment of Confederate landmines actually killed and injured a number of Union soldiers at Yorktown, Virginia, on May 4, 1862, did the infernal machine become a major feature of the Civil War.

LEXINGTON

The siege of Lexington was part of the early days of the war in Missouri. This border slave state was deeply divided in sympathies, but a small Federal army largely secured the Missouri River valley in June 1861 and drove into the southwestern part of the state later that summer. It was defeated at the battle of Wilson's Creek near Springfield on August 10 by a combined force of Confederate troops from Arkansas, Louisiana, and Texas in alliance with the pro-Confederate Missouri State Guard under Maj. Gen. Sterling Price. After Wilson's Creek, Price broke away from his Confederate colleagues and moved toward a Union garrison at Lexington fifty miles east of Kansas City on the south bank of the Missouri River.[1]

Lexington was held by a force of 3,500 Federals under Col. James A. Mulligan of the Twenty-third Illinois. On the morning of September 11, the troops started to construct earthworks enclosing an area of eighteen acres. In a lecture given by Mulligan before he died in late July 1864, the Federal commander referred to "confusion pits" and "mines" placed before the defenses. Exactly what he meant by either term is an open question. Another description of the defenses by Capt. Joseph A. Wilson refers to *trous de loop* and wires but does not refer to mines or torpedoes. William C. Crowell of Company A, Twenty-third Illinois, in a postwar account, also refers to "mines" when describing the defenses but in a casual, offhand way that reflects the probability he simply read Mulligan's equally casual, offhand reference to them. There is no reliable evidence that anything like a landmine was planted at Lexington, despite Mulligan's and Crowell's references to the device.[2]

Price appeared on September 12, barely twenty-four hours after the start of fortifying, with an army of ten thousand men. During the next week, these pro-Confederates cut off the Federals south of the Missouri River and advanced under cover of hemp bales until they hemmed in the garrison so closely that Mulligan surrendered on September 20. Price evacuated Lexington soon after to avoid other Union columns.[3]

PORT ROYAL

In contrast to Lexington, there is no doubt that the Confederates planted at least one explosive device before evacuating their position at Port Royal, South Carolina. Targeted by a Union army and navy expedition to obtain possession of an important sound and associated port facilities to support the Federal blockade, the Confederates had constructed fortifications filled with heavy artillery to contend with attacking vessels. The Federal fleet under Flag Officer Samuel F. DuPont executed a smartly-planned assault on November 7, 1861. It managed to overpower both Fort Beauregard on Phillips Island on the north side of Port Royal Sound and Fort Walker on the north end of Hilton Head Island on the south side. Federal troops landed to take possession of the area on November 8.[4]

Company D, Third Rhode Island Heavy Artillery entered Fort Beauregard, but some alert men discovered a powder train leading to its magazine. A device had been erected so that when the Confederate flag was lowered, it "would at the same instant fire the diabolical train," in the words of the regimental historian. The trick did not work; no one was hurt. This torpedo was in the form of a booby trap designed to catch an unwary man with potentially deadly results.⁵

But another booby trap planted near Fort Beauregard was touched off. Capt. Daniel Ammen, commander of USS *Seneca*, went ashore and saw a small frame house located fifty yards from the fort. He entered briefly and left again. "In a few minutes an explosion was heard, and, on turning, I saw a cloud of smoke where the house had stood." An explosive device had been planted beneath the building and had been touched off when a sailor passing by hit a wire with his foot. The sailor "was knocked over and partially stunned, but soon revived." It had been a close call for Ammen and especially for the sailor.⁶

After the war, while recounting his experience, Ammen criticized the enemy for planting these two booby traps. "It may be said that it is natural in warfare to harm your enemy as much as possible, but it strikes the man who has escaped being blown up that such devices are essentially mean." It was typical for these early Civil War instances of torpedo warfare that no evidence can be found in Confederate sources concerning them. Rebel engineer Maj. Francis D. Lee filed a report of the Port Royal engagement but made no mention of booby traps or torpedoes of any kind. We do not know with certainty if they were placed with Lee's knowledge, but one would assume he could hardly fail to know of them. There likely was a conspiracy of silence concerning these devices at Port Royal as there would be later at Yorktown.⁷

COLUMBUS

The Confederates deployed the first major system of landmines in American history as part of their overall plan for the defense of Columbus, Kentucky. Troops under Lt. Gen. Leonidas Polk occupied this town situated on high bluffs bordering the east side of the Mississippi River, twenty-five miles downstream from the Union stronghold of Cairo, Illinois, in early September 1861. Columbus was the western end of the Confederate defensive posture that stretched across southern Kentucky, dipping down into Tennessee to include Fort Henry on the Tennessee River and Fort Donelson on the Cumberland River. That posture continued with the fortifications protecting Bowling Green, Kentucky, and went east to include Mill Springs and, finally, Cumberland Gap. Although fortifications existed at all these locations, none of them were as extensive as at Columbus.⁸

Polk had several engineer officers at Columbus and received full support from others in his plan for the defense of the place. Matthew Fontaine Maury crafted underwater torpedoes fired by electricity and offered up to twenty-five of them to Polk on December 4, 1861. He also wrote detailed instructions on how to place them in the Mississippi River. But Polk had closer sources of underwater mines. A. L. Saunders at Memphis wrote to him on December 5 about his own construction of "improved submarine batteries," nineteen of which were sent by rail to Columbus as part of a total order of fifty. Saunders also wrote instructions about how to place the devices so they would not interfere with the passage of Confederate vessels. They were to be charged with twenty-five pounds of gunpowder just before submerging, but the levers which activated the charge were not to be inserted until the mine had been properly weighted and positioned. Until the levers went in, the charge was virtually harmless. It is possible that not all fifty of these devices were sent to Columbus before its fall. In July 1862 Confederate operatives still possessed seven of them and were ready to send the torpedoes to Vicksburg during the first Union offensive against that Mississippi River town.[9]

"I am paving the bottom of the river with submarine batteries," Polk exulted to his wife on January 6, 1862. He thought they would "take good care . . . of Mr Lincoln's Gun Boats." But he was moving forward with planting "fougasses or mines out on the roads also, so that if they make their appearance, we will not fail to give them a warm reception." The Federals became aware of these preparations. Brig. Gen. Ulysses S. Grant received reports of the obstructions at Columbus as early as January 6, 1862. The Confederate scheme included a heavy chain stretched across the river with contact torpedoes attached to it. Grant's informant indicated that the Confederates had tried one of them by pushing a coal barge against it and the device detonated as expected. While this information was pretty accurate, Grant received further reports that were not reliable. He had been told that more torpedoes were distributed in the river upstream from Columbus and that six hundred devices had been deployed south of the town, all the way down to Memphis.[10]

It is interesting that Grant had no news about the planting of landmines at Columbus. The Federals and their informants were only concerned with the underwater explosives. Thus, the landmines were a big surprise when the Federals occupied Columbus without a battle. Grant's capture of Fort Henry on February 6 and Fort Donelson on February 16 broke open the Confederate defensive posture in southern Kentucky. Rebel forces evacuated Bowling Green, and Polk began to evacuate Columbus on February 25, shipping his sick men and his commissary, quartermaster, and ordnance stores south. He moved most of his troops by March 1, keeping a rear guard in position until Federal cavalry approached on March 3.[11]

Those Union troopers were Lt. Col. Harvey Hogg's battalion of the Second Illinois Cavalry. Hogg was supported by five gunboats commanded by Flag Officer Andrew H. Foote, escorting a small infantry force down the Mississippi, which landed at Columbus on March 4. "The flag of the Union is flying over the boasted 'Gibraltar of the West,'" reported Maj. Gen. Henry W. Halleck, commander of Union forces in the Mississippi Valley.[12]

Halleck's chief of staff, Brig. Gen. George W. Cullum, was with Foote, and as soon as he landed, he discovered smoke issuing from both ends of the large Confederate magazine on the bluffs. He instructed some soldiers to cut the powder train leading to both ends before sparks communicated to the main store of gunpowder. Obviously, the Confederates had arranged to blow up their store upon evacuation. Whether this was meant to go off after the Federals occupied the place or something went wrong with their plans and both powder trains merely smoldered is not known, but Cullum's alertness saved many lives.[13]

Foote also saw a large pile of underwater torpedoes left behind by Polk's engineers on the river bottomland. Artist Henri Lovie arrived at Columbus on the evening of March 5 and drew a sketch of this pile that was published as a woodcut illustration in *Frank Leslie's Illustrated Newspaper* on March 29, 1862, giving the Northern people their first look at Confederate underwater torpedoes. Lovie wrote of his impressions in a letter dated March 8, which also was published in the newspaper. He referred to "a huge pile of torpedoes and infernal machines which the rebels had left on the shore, some still in boxes, addressed to Gen. Polk, Columbus. Weights, anchors and iron rings were intermingled with the torpedoes."[14]

Without realizing it, Lovie hinted at the existence of the landmines when he referred to a telegraph office dug into the bluffs with copper wires extending to other parts of the defenses. But it was up to Capt. William A. Schmitt of Company A, Twenty-seventh Illinois to discover what this meant. For two days he and his men explored the Confederate defenses, intrigued by "ridges of new earth, similar to ridges which are formed by covering up gas or water pipes in a city," according to a correspondent of the *Chicago Times*. Inside the cave mentioned by Lovie, they found an "implement similar to those used in a telegraph office, with wires running in a dozen different directions." By following the raised ridges, the Illinois troops reportedly discovered seventy-five to one hundred torpedoes planted along the north and northeast perimeter of the Confederate defenses.[15]

Various people offered differing descriptions of these landmines. The *Chicago Times* correspondent referred to one as "a large iron cask," which was three feet tall and one and a half feet wide, "in shape as near as can be described to a well-formed pear, with an iron cap fastened by eight screws." When Schmitt's men removed the cap, they found four eight-pound shells with loose grape and canister and two bushels of gunpowder inside the cask. In the ground and below the cask, they also found "several batteries, with hollow wires attached to two larger wires, covered with a substance impervious

Figure 2.1. Columbus Torpedo. The Confederates deployed about a dozen powder chambers like this one, which looks like a baking pot, at locations along two roads leading to their fortified position at Columbus, Kentucky. They rigged both concentrations of landmines to be detonated by electricity but blew none of the charges before evacuating Columbus. (Lossing, *Pictorial Field Book*, vol. 2, 237)

to water, connecting to the cavern." The Federals uncovered about a dozen such iron pots, each one occupying a pit five feet deep.[16]

The *Scientific American* published a letter by an anonymous writer who examined the landmines at Columbus in its April 5, 1862, issue, further spreading word of the system to the public. He described the torpedo as "shaped like an old-fashioned tea kettle in some respects, and has a sort of cap on the top similar to the lid of an iron tea kettle, which is fashioned to it by a set of iron screws. About two inches from the bottom are two orifices through which run copper wires, insulated with gutta percha and tarred cord, which are laid in trenches and communicate with a galvanic battery inside of the fortification on the bluff." In fact, this anonymous correspondent offered the most detailed and reliable description of the landmines at Columbus. He was even more detailed and scientific when describing the underwater mines left behind by the Confederates.[17]

Lovie correctly noted that the detonating machine was similar to those that powered telegraph lines, a fact noticed by other observers too. The *Scientific American* letter confirmed that it was a galvanic battery, or an electrochemical machine, likely the one developed by Grove because that type was the most commonly used by telegraph companies in the United States. Historians Michael P. Kochan and John C. Wideman assume the operator used coils so they could have "lower voltage in the system to maintain continuity, but jumped the voltage up to fire the mines when required." Apparently, they used "common telegraph keys" to operate the system.[18]

Cullum directed a topographical survey of the defenses at Columbus to produce a detailed map, which included the landmine system. The map identified two battery rooms for the detonation of torpedoes. One of them was located in the outer ditch of a large arrowhead fortification almost due east of Columbus. Six wires extended from there to torpedoes planted on the Clinton Road, which approached the work from the east. The other battery room was located in a sunken pit north of Columbus, with wires connecting to torpedoes on Elliott's Mill Road, which ran along the top of the bluff nearly parallel to the Mississippi River.[19]

Word of these landmines spread rapidly through the public media and through word of mouth among soldiers and civilians, although accurate information sometimes was at a premium. Edward W. Crippen of the Twenty-seventh Illinois, who was on the scene, incorrectly noted in his diary that the Confederates had planted torpedoes "promiscuously as far back in the country and as far up & down the river as the fortifications extend Each one connecting by wire with a magnetic Battery by which they were to be exploded."[20]

In fact, Cullum's survey documented only two battery rooms with six wires leading from one and an unreported number of wires extending from the other. Cullum's investigation also implies that there could not have been as many as one hundred torpedoes. If we base our conclusions on his map, there were no more than six along the Clinton Road and probably no more than the same number along Elliott's Mill Road, only a dozen altogether.

It is true that the cask which appears in all detailed descriptions of the torpedoes was specially designed and manufactured. Apertures were placed exactly where they were needed and would have been useless if the cask was originally designed to hold liquid. Just as Saunders specially made his underwater torpedoes, he might have contracted with someone to make these casks to order. The landmine system at Columbus, in other words, was not improvised.

The Columbus landmine system marked a significant milestone. It was not only the first landmine system deployed in American history but was well planned with an electrochemical detonation device. The torpedoes were designed to harass Union troop movement close to the earthworks along two roads the Confederates assumed their opponent would use in approaching Columbus, one coming from Paducah to the northeast and the other coming from Fort Holt in northwest Kentucky. The Confederates specially designed and manufactured the torpedoes for the purpose rather than employing artillery shells as Rains had done at Fort King twenty-two years before.

Northern reaction to the report of torpedoes at Columbus was caustic. The editor of *Frank Leslie's Illustrated Newspaper* spared nothing in condemning the Confederates. He linked the deployment of landmines at Columbus with other reports of Southern barbarity, such as poisoned wells and food. "The rebels seem by instinct, which really we must term unnatural, to have collected all that is really cowardly, treacherous, base and brutal together in one code, and made it their entire tactics. . . . Whatever is repulsive and abhorrent to true manhood," the editor continued, "finds its congenial home in the bosom of a Southern rebel."[21]

Nine months after the fall of Columbus, Thomas D. Beggs of the 114th Illinois was on his way to rejoin his regiment when he stopped briefly at Columbus in December 1862. The chain that had been stretched across the Mississippi had long since been taken up but was still piled on the bottomland. Beggs called it "Pillow's Watch Chain," and recalled reports of torpedoes, which had earlier made such headlines. He thought all such contrivances were "good monuments of secesh folly and foolhardiness."[22]

It was easy enough for Beggs to refer to the Columbus torpedoes as examples of Confederate folly because none of them went off and hurt anyone. But if they had exploded in action and killed or wounded Federal troops, that would have elevated their use into another level of moral outrage.

Chapter 3

Yorktown

"The Horrible Realities of War"

The Yorktown phase of the Peninsula Campaign, which lasted from early April until early May 1862, brought the use of landmines fully to the attention of people in the North and South. The Confederates deployed an improvised system using artillery shells and friction primers in and around their earthworks across the Yorktown Peninsula before evacuating the position. Several Federal soldiers were killed or injured in what became the most widely known example of torpedo use on land in the Civil War. This exposure brought into play discussions about the morality and the tactical effectiveness of landmines to its fullest extent thus far in the conflict.

Although he refused to admit it, Gabriel James Rains was responsible for those landmines at Yorktown. Following his duty in the Second Seminole War, Rains saw service in Texas in the early days of the Mexican War, participating in the battles of Palo Alto and Resaca de la Palma. Then he was sent off to recruiting duty and received a promotion to major after the war. Rains spent several years in Washington Territory and California before his promotion to lieutenant colonel of the Fifth US Infantry and service in Vermont, just before the outbreak of the Civil War. A staunch Southerner, he resigned his commission in the US Army on July 31, 1861, after thirty-four years.[1]

With his lengthy service and high rank in the prewar military, Rains was appointed a brigadier general in the Confederate army on September 14, 1861, and placed in command of a brigade consisting of Georgia, Louisiana, and North Carolina regiments. Since at least early October 1861, his brigade was assigned to the Yorktown Peninsula. At times he temporarily commanded a division and was in charge of the post of Yorktown. He performed well enough to encourage his immediate superior, Maj. Gen. John B. Magruder, to recommend that Rains be promoted to major general.[2]

The Federals arrived on the Yorktown Peninsula when Maj. Gen. George B. McClellan brought the Army of the Potomac in early April and planted it opposite the line of earthworks stretching across the neck of land from Yorktown. Gen. Joseph E. Johnston's Army of Northern Virginia held the position with its left wing, under Maj. Gen. Daniel Harvey Hill, holding trenches from Yorktown partway across the peninsula. Rains's Brigade was an element of Hill's command. The Federals dug their own trenches and constructed emplacements for heavy artillery, planning to bombard the Confederate works into submission. But Johnston evacuated the Yorktown line on the night of May 3, just before the Union guns were ready to open fire.[3]

What happened next became a bitter controversy. Rains refused to admit any responsibility for the landmines encountered by Federal troops, but there is overwhelming evidence that he planted them. A few days after Yorktown's evacuation, a man named Grover from western New York state, who had served in the Confederate army since the beginning of the war, either deserted or was captured by Federal forces. He told them that Rains had been responsible for the planting of numerous torpedoes at Yorktown. Grover claimed that he personally saw Rains superintending the placement, assisted by a man named Gray. They had a wagonload of torpedoes and put them "in particular spots" at several locations along the line.[4]

Maj. James W. Ratchford, Hill's adjutant general, also testified that Rains was responsible for the torpedoes at Yorktown. Rains operated with a prearranged plan. He planted artillery shells in front of selected sectors of the works with a cap over the fuse plug so that he could place a friction fuse in the plug at the last minute to reduce the possibility that the Confederates might be hurt by them. When the order to evacuate was circulated, Rains fused those shells and quickly planted more near abandoned stores to further harass and injure the Federals. "I think that all Gen. Rains did about the torpedoes was on his own responsibility," Ratchford stated after the war.[5]

"Reliable hands were employed burying 'tor-paddos' as an Irishman called them, all about inside our camp," wrote Lt. Robert T. Hubard Jr., a Virginia cavalryman, on May 2. Federal troops who saw these devices agreed that they were numerous and spread out. "Many were in groups, others planted at haphazard," recalled Pvt. Robert Knox Sneden, a draftsman at Third Corps headquarters. "In fact the whole place is mined," stated a member of the Fifth Massachusetts Battery with some exaggeration. They were planted near earthworks, on roads, and on the streets of Yorktown. Explosive devices appeared "in any place that was likely to be visited by our men, on the walks by the forts and between the graves where rebel soldiers had been buried," wrote a Federal soldier.[6]

Unaware of the danger, many Unionists who entered the Confederate position became the first American victims of landmines on the rainy morning of May 4, 1862. On the sector held by Brig. Gen. Fitz John Porter's First Division, Third Corps, two companies of the Sixty-second Pennsylvania and two companies of the Twenty-second Massachusetts moved into the Confederate works formerly held by Hill's Wing at 5:30 a.m. On the move into the works, someone tripped a torpedo, and six men of Company G, Twenty-second Massachusetts were wounded. Col. Jesse A. Gove referred to "these inhuman missiles of war" when reporting the loss in his regiment. The other Massachusetts men learned quickly from this gruesome experience. They carefully looked at the ground and avoided any sign of disturbance on the surface. They also had the good sense not to walk through the gate of the major fort on this part of the line, fearful that a mine was planted there.[7]

Trouble developed outside the fortified area as well. The McClellan Dragoons (Company H and Company I, Twelfth Illinois Cavalry) were moving along a road approaching the Confederate line and ran onto torpedoes. One of them exploded and wounded two horses and a trooper. A short distance away, the Forty-fourth New York relieved the Twenty-second Massachusetts at 2:00 p.m. on May 4, and the Massachusetts men marched to their camp south of the Confederate works. On the way Wallace H. Gilbert of Company F "stepped on a buried torpedo, exploding the cap without bursting the shell," wrote the regimental historian. "It was a very narrow escape." Another man belonging to Company C saw what he assumed was the fuse of a torpedo sticking above the road surface. He "gave it a kick, and threw himself flat," but there was no explosion. On further investigation he found it was "a tumble-bug, whose shiny black body" resembled the top of a fuse.[8]

Another regiment in Brig. Gen. Charles S. Hamilton's Third Division of the Third Corps experienced a brutal introduction to torpedo warfare. The Fortieth New York marched into a Confederate fort at 6:00 a.m., and Col. Edward J. Riley shouted, "Order arms." When the men thumped their musket butts on the ground, Pvt. George McFarrar set off a landmine. McFarrar and Pvt. Michael McDermott were killed, and three others were wounded, one of whom later died. According to Asst. Surg. Charles E. Halsey, the lower limb of one of the victims was thrown thirty feet from his body. Riley immediately ordered the regiment out of the fort and posted a guard to stop anyone from entering it.[9]

All this had taken place near the left end of the Confederate line, but trouble also developed along the right wing of the Rebel position. There elements of Brig. Gen. Silas Casey's Third Division, Fourth Corps moved along a road leading to the enemy works, when they encountered several torpedoes. One of them exploded, killing one man and injuring six others belonging to the Fifty-second Pennsylvania. Members of the Eighty-fifth Pennsylvania, moving farther back in the column, heard the "stunning report" of the explosion and were forewarned. When the regiment came to the scene, Lt. Col. Harry A. Purviance "found the road torn up, and a wounded soldier lying at the roots of a tree near by. One of his comrades had been killed.... We had a sad foretaste of what the devils had prepared for us." Quickly taking in what had happened, the Eighty-fifth Pennsylvania stopped as Purviance and other officers told the men to carefully scan every inch of the road before moving about.[10]

The Sixty-ninth Pennsylvania in Brig. Gen. John Sedgwick's Second Division, Second Corps suffered a gruesome casualty about an hour after it occupied a Confederate fort. Pvt. John Greene of Company D stepped on a torpedo. "His left leg was torn off at the knee and carried over the immense rampart into the ditch" of the work. Surg. Charles E. Bombaugh of the Sixty-ninth Pennsylvania found him "still conscious, but sinking from hemorrhage and nervous shock." Bombaugh cut off part of the remaining thigh at the middle and tied the femoral artery, but Greene stopped breathing before he could finish his work. The surgeon never forgot the moment of Greene's death. "It was a sorrowful scene, and one made more solemn by the gloom of the drizzling rain, and more impressive by the indignant faces and the muttered vengeance of the hundreds who were looking on."[11]

The Federals entered a different phase of their introduction to torpedo warfare when they realized that the village of Yorktown and some areas outside it were laced with booby traps. "You could not tip over a barrel, or anything else, but what had a string attached to a big shell or some kind of torpedoes," wrote Peleg W. Blake of the Fifth Massachusetts Battery. "Wherever you could see the dirt thrown up loosely, look out for your feet." An overcoat lying on the ground had a string attaching it to an explosive device. In one of the Yorktown houses, a coffeepot served the same purpose, and at another location, it was a bag of flour. Many soldiers heard about a New York soldier who saw a pocket knife on the ground and was killed when he tried to pick it up. Chaplain James J. Marks of the Sixty-third Pennsylvania witnessed this sad incident.[12]

The most visible victim of a torpedo at Yorktown was not a soldier but a civilian telegraph operator working for the government. D. B. Lathrop hailed from Springfield, Ohio, and had studied for the ministry before choosing instead to learn telegraphy. Assigned to the staff of Maj. Gen. Samuel P. Heintzelman, commander of the Third Corps, Lathrop was eager to see if the Confederates had left behind a functioning line in Yorktown. He located the office of the Rebel telegraph operator and was removing a ground wire when he stepped onto a torpedo hidden beneath the floor. The explosion tore off one or both legs, according to differing reports, and he died soon afterward.[13]

Many Federals became aware of their danger by hearing the explosions touched off by their unfortunate comrades. "I see a good many of the Torpedoes that they laid for us," wrote Moses Hill of Andrew's Sharpshooters, a unit attached to the Twenty-second Massachusetts. "But I was sure not to go very close to them. I see one go off and it blowed one man most all to peaces."[14]

Figure 3.1. Yorktown. This dramatic illustration, based on a drawing by Alfred R. Waud, captures the shock, horror, and destructive potential of Confederate landmine explosions at Yorktown. (*Harper's Weekly*, May 24, 1862)

Heintzelman, whose troops were more affected by the mines than those of other commands, ordered guards to be placed until thorough searches could be conducted for these devices. He rode to the entrance of one fort but was stopped by a sentinel who told him the gate had been mined. Heintzelman and some staff members therefore clambered through the ditch and up the parapet, where they observed what they could of the earthwork. Maj. Charles S. Wainwright, artillery chief of Brig. Gen. Joseph Hooker's Second Division, Third Corps, was moving his guns along the road skirting the fort Heintzelman stood on. As Wainwright rode closer to take a look at the work, Heintzelman warned him off, yelling that the fort was mined and he should remain on the road. That was not necessarily a safe place either; Wainwright saw about a dozen spots on the roadway, which the Federals had already marked as torpedo plantings. "All that was visible was a foot or two of telegraph wire sticking out of the ground," ready to ensnare a foot or hoof and trip the explosive.[15]

The torpedoes Wainwright saw probably were marked by sticks with red or white rags tied to them, a tactic employed by the Federals at many locations that morning. A sergeant in the Tenth US Infantry reported that they also forced captured Confederates to locate and mark some torpedoes. Lt. Charles A. Phillips of the Fifth Massachusetts Battery saw a "bloodstain on the ground where a man was blown up by one of the rebel infernal machines, and a little red flag about ten feet from it."[16]

Warnings spread quickly through the ranks. As Brig. Gen. Oliver O. Howard's First Brigade, First Division, Second Corps began to move out of its camps, word came along the moving column to be aware of "buried bombs." The road was covered with thick, gooey mud several inches deep, and the column moved slowly. Howard stopped where the head of his command passed through the Confederate line of works and saw many small flags marking the location of known torpedoes in the area. He and some staff members remained there and called out to the men not to stray from the road and to avoid any flagged spots on the roadway itself. They became hoarse with the yelling because the brigade moved slowly and took nearly all day to pass the danger spot.[17]

The Federals only needed a warning; most of them were only too ready to protect themselves from danger. As Battery B, First Rhode Island Light Artillery moved through the Rebel works, its members caught glimpses of barrels and boxes with the word "Danger!" written on them. They knew what it meant when word circulated that some men in other units had been killed or injured while searching for and marking the locations of torpedoes.[18]

There were many close calls. Surg. Charles E. Bombaugh of the Sixty-ninth Pennsylvania rode into a Confederate fort to look around and learned on leaving it that some torpedoes had been discovered there. "Either the pouring rain had melted the fuses, or my lucky horse chanced to step on harmless ground," he mused. On the far Union left near Lee's Mill, Brig. Gen. John W. Davidson's Third Brigade, Second Division, Fourth Corps set out from its camp. Col. Edwin C. Mason of the Seventh Maine had already heard of the fate of D. C. Lathrop, an acquaintance, so he was forewarned. Mason moved ahead of his regiment and suddenly "heard beneath his foot the cracking of a percussion cap," in the words of Oliver O. Howard, who undoubtedly heard the story from Mason. The Maine officer bent over and carefully moved the mud about to find a red wax capping on the fuse of a shell. He had snapped the percussion primer, but the shell had not detonated. Mason then called for volunteers from his regiment and set them to work on their hands and knees to find other torpedoes in the area.[19]

According to an anonymous correspondent of the *Cincinnati Commercial*, McClellan himself narrowly avoided injury or death by one of these devices. He was riding through the Confederate line in an ambulance where the roadway was blocked. The driver started to leave the road to the right but came to an obstruction and stopped to remove it. The driver found "a bed of torpedoes almost under the wheels of the vehicle. Had he driven a few feet further the Commander in Chief might have been blown to atoms." This correspondent also reported accurately about the death of John Greene of the Sixty-ninth Pennsylvania, so his report about McClellan needed to be taken seriously.[20]

Even after passing the most dangerous places at and near the Confederate works, the Federals continued to encounter danger on their march toward Richmond. Torpedoes appeared along the road to Williamsburg, the primary line of Confederate retreat and Union advance. Flags appeared at every location found, and where two or three torpedoes were clustered, a sentinel was placed to provide additional warning to passing troops.[21]

Third Corps units moving north along the Williamsburg Road came upon "heaps of fresh earth at intervals" in a field on both sides of the roadway just north of Windmill Creek at about 9:00 p.m. They became suspicious and everyone stopped until the mounds could be investigated. Artillery men took the lead because they handled ordnance as a matter of course. Carefully digging into several of these mounds, they found nothing. It is possible, as a draftsman at Third Corps headquarters asserted, that the Confederates deliberately created these mounds to confuse their enemy, but they could certainly have been created by some civilian for other purposes. At any rate, the column was delayed no more than an hour by this incident.[22]

McClellan's pursuing army caught up with the rear guard of Johnston's command near Williamsburg, where a sharp battle took place on May 5 with inconclusive results. The Confederates continued to retreat and Rains's Brigade was assigned to cover the rear. He decided to do something to hinder pursuit. On May 6, Rains came upon a broken-down caisson stuck in a mud hole in the road. There were five shells in it, so he gave them to five of his men to carry. Rains coordinated his work with the cavalry commander who was cooperating with his brigade and placed four of the shells in the road "a little beyond a fallen tree." Carefully supervising the placement, he put three of them in a cluster about a yard apart in triangular formation and the fourth "a little to the left in a basket." He did not report what was done with the fifth shell, but Rains wrote after the war that he "happened to have" some primers with him, inserted them into the four shells after they were buried, and covered them till out of sight.[23]

Not long after leaving the spot, Rains heard two explosions and assumed pursuing Union cavalry had gotten to the area of his planting. He believed the three shells planted in triangular formation had all exploded and the fourth one went off separately. After the war, Rains claimed that this surprise so stunned the Federals that it halted the pursuit by McClellan's army for three days.[24]

There is no doubt that Rains planted these four torpedoes at some spot on the Williamsburg Road some distance from Williamsburg on the morning of May 6, and it is possible they were touched off by pursuing Federals. But his claim that it caused the entire Army of the Potomac to allow Johnston to continue retreating cannot be taken seriously. There are no reports of any kind by Federal soldiers regarding this incident, and McClellan's troops continued moving forward without delay.

Troops assigned to stay at Yorktown for a while conducted themselves with extreme caution for several days after the evacuation. On May 7, Surgeon Bombaugh of the Sixty-ninth Pennsylvania reported that "we step as if on glass, or on the brink of a precipice, every now and then discovering one of the little red fuses thickly planted by rebel barbarity."[25]

It is not known how long it took details to remove all the torpedoes at Yorktown, but large numbers of men were set to the task. Two companies of the Fiftieth New York Engineers were detailed to look for and dig up landmines. Draftsman Robert Knox Sneden reported that many torpedoes were dug up by the engineers in only two hours' time, but because others were so scattered, it would take much longer to find them.[26]

But an unnamed Union officer added some numbers to the story of mine clearance when he told a correspondent of the *New York Herald* that thirty devices had been discovered within an area of ten square rods. All of them were eleven-inch artillery shells. When Capt. William B. Weeden of Battery C, First Rhode Island Light Artillery looked into one of these eleven-inch shells, he found close to four pounds of "very coarse and very fine powder mixed." The correspondent continued to describe what Weeden found. "Each has a quill fuse, and above it a plunger, with knob so constructed that a person walking along and stepping upon it brings the plunger down with sufficient force upon a cap underneath to cause it to explode."[27]

A number of men were sent out to conduct a wide sweep of the area, including the streets of Yorktown, deserted Confederate camps, roadways, and even adjacent fields. They did a thorough job which probably uncovered virtually all the torpedoes the Confederates had planted, although the possibility that some would not be discovered until much later always existed. To a limited degree, the Federals employed captured Confederates to locate and dig them up, although there are few details concerning this policy. It would have been faster and more reliable to use Union military personnel, especially engineers and artillery men, and that appears to have been the main force used in mine clearance at Yorktown.[28]

McClellan reported that "some losses" occurred from these torpedoes, but word concerning the danger spread rapidly enough "to prevent many injuries." It was commonly repeated that about a dozen Federals were killed or injured by them on May 4, 1862. The Confederates stated their belief that the number was about thirty, but their sources of information could not have been as reliable as those employed by Federal soldiers.[29]

Moral outrage immediately characterized Northern reaction to the Yorktown torpedoes. McClellan reported the incident to Sec. of War Edwin M. Stanton at 7:00 p.m. on May 4. "The rebels have been guilty of the most murderous & barbarous conduct in placing torpedoes *within* the abandoned works, near wells & springs, near flag staffs, magazines, telegraph offices, in carpet bags, barrels of flour, etc." Losses amounted to four or five killed and about a dozen injured. The general intended to force Confederate prisoners to take them up. To his wife, McClellan referred to the Rebels as villains for scattering mines everywhere. "It is the most murderous & barbarous thing I ever heard of," he concluded.[30]

Literally every Federal soldier who commented on the incident agreed with McClellan. Some called the Confederates "a very mean set of people," while others expounded more vigorously on the character of their enemy. George R. Buckman of the First Minnesota termed them "the most inhuman and barbarous class of beings that ever lived in any country." Calvin D. Mehaffy, a regular army officer on the staff of Fitz John Porter, acknowledged that it was legitimate to plant such devices before a position to impede the opponent's ability to capture it. But to evacuate the position and scatter torpedoes about to kill individuals was "downright murder." "If in so doing they could kill fifty thousand men it would be perfectly fair and not inconsistent with warfare, but to blow up a telegraph boy in the pursuit of his calling is both murderous & barbarous."[31]

Northern newspapers gave full coverage to the Yorktown torpedoes and the casualties they caused. The *New York Herald* printed McClellan's report to Stanton, and other papers printed letters from soldiers who uniformly condemned the Confederates. Many editors chimed in with condemnations of their own. "War, at best, is horrible enough," wrote the editor of *The Alleghanian* of Ebensburg, Pennsylvania. "But the unnecessary slaughter of human beings against the rules of warfare which all civilized nations respect, or to gratify a mere blood-thirsty spirit of revenge, without aiming at or expecting any military results, is as essentially sheer murder during the existence of hostilities as it would be in times of profound peace."[32]

By this stage of the war, a year into the bloody conflict, a number of incidents of what Northerners considered barbaric conduct by their enemy had accumulated. The planting of torpedoes at Yorktown was quickly added to the list. "They made drinking cups, and drum sticks, rings & c of the bones of the men who fell at Bull Run," asserted George R. Buckman, who claimed to have seen the body of a Federal soldier with his throat cut after having been wounded. The editor of the *Nashville Union*, a Federal newspaper published after the city fell to Union troops, reeled off a list of atrocities that included firing on flags of truce, shelling hospitals, killing Unionist men, and driving women and children out of east Tennessee, as well as now hiding explosive devices to kill unsuspecting soldiers. "Sin always arrogates to itself the widest license," the editor concluded.[33]

Lt. William Byrnes of the First Michigan reported that booby traps had also been planted at Gosport Navy Yard. When the Confederates abandoned the facility early in May, Federal troops found a few shells rigged up to explode. Apparently, no one was hurt by them. Byrnes was not happy about it. "This is honorable warfare, this is having [to] deal with Chivalry, this their boasted magnaminity—Bah."[34]

The Yorktown torpedoes had a long life in the memory of Northerners. They were mentioned in unit histories, memoirs, and reminiscences with words such as "villainous," "savage," "brutal," and "cowardly." For those most directly affected, the landmines "brought the horrible realities of war home to us for the first time," wrote Fred C. Floyd, historian of the Fortieth New York. That regiment suffered more torpedo casualties than any other at Yorktown. The event created a spirit of revenge. Floyd wrote that "the severity of discipline alone prevented the exaction of a penalty on the spot," implying that some New Yorkers wanted to harm Confederate prisoners but were stopped by their officers. Thomas W. Hyde of the Seventh Maine also recalled that the torpedo incident was a turning point in emotional attitudes toward the Confederates. "We had not had the time to get up much of a feeling of hostility to them as yet."[35]

Anger and resentment about the Yorktown torpedoes would never die in the North until the passing of the Civil War generation. Those who condemned the Confederates for planting them were primed by their pre–Civil

War attitude toward secret explosive devices. Newspaper reports about torpedo crimes had conditioned them to view secreted landmines, especially booby traps, as attempted murder.

Ironically, a handful of Rebel soldiers became victims of these torpedoes at Yorktown. Maj. Edward Porter Alexander, chief of artillery on Maj. Gen. James Longstreet's staff, recalled that an explosion had been heard about 3:00 a.m. of May 4 and learned later that "some cavalry stragglers from the Confederate rearguard, who entered the town," had stumbled upon a booby trap. Some troopers were wounded. Lt. Robert T. Hubard Jr. of the Third Virginia Cavalry reported that these stragglers tried to break into a house they thought was filled with flour, but it actually was rigged so that opening the door would set off a shell. Many other shells were stored in the house as well so that when the Rebels opened the door and suffered wounds for their trouble, the house also went up in flames. Shells exploded for some time after, and the fire cast a lurid glare on the streets of Yorktown during the dark hours of the night.[36]

Confederate views of these torpedoes at Yorktown are scarce. There is virtually no reaction to them in the personal accounts by common soldiers, but high-ranking officers and their staff members noticed and commented on the landmines. Their reaction was initially one of disdain on the part of some men and support on the part of others.

But a serious reaction to the torpedoes among the Confederates did not set in until May 11, a week after the event. Johnston reorganized his army structure, placing Hill's Division in a new Second Corps commanded by Longstreet. In giving instructions for taking position near Richmond that day, Longstreet remembered the report that Rains had planted torpedoes at Yorktown and did not want it to happen again. Capt. G. Moxley Sorrel, his assistant adjutant general, informed Rains of this concern. "It is the desire of the major-general commanding that you put out no shells or torpedoes behind you, as he does not recognize it as a proper or effective method of war."[37]

Sorrel's language in this May 11 dispatch was guarded but explicit. In his memoirs, however, the staff officer was more dismissive. Rains had "amused himself planting shells and other explosives in the roadway after us to tickle the pursuers," he wrote. Sorrel was referring in his memoirs only to the planting of shells on the Williamsburg Road on May 6. Furthermore, he argued that he had informed Longstreet of that incident immediately after May 6, and the general "instantly stopped" further planting. That cannot be true, considering that Rains was not even in Longstreet's command at this time. Yet Sorrel claimed in his memoirs that Longstreet instructed him to write "a rather severe note" to Rains "reminding him that such practices were not considered in the limits of legitimate warfare, and that if he would put them aside and pay some attention to his brigade his march would be better and his stragglers not so numerous." There is no evidence to support Sorrel's claim that Longstreet reprimanded Rains between May 6 and May 11. It is more likely that the staff member referred in his memoirs to the mildly worded dispatch of May 11 than to a severely worded reprimand written earlier than that date.[38]

Another reason to doubt that a severe reprimand was written before May 11 is that, even when reading the mildly worded dispatch of May 11, Rains vigorously asserted that what he had done was right. If a worse scolding had been delivered before that date, he most likely would have let everyone know his feelings about it. "Believing as I do the vast advantages to our country to be gained from this invention," he wrote in an endorsement on the May 11 dispatch, "I am unwilling to forgo it, and beg leave to appeal direct to the War Department."[39]

For the first time in his contentious life, Rains developed a larger theory concerning landmine warfare in order to convince his superiors of its morality and tactical effectiveness. Continuing in his endorsement, he placed landmine use on the same level as setting an ambush for the enemy, erecting masked batteries of artillery, and using underground galleries to blow sections of enemy earthworks into the sky. He repeated a rumor that McClellan had authorized the undermining of Confederate Redoubt No. 4 at Yorktown. "If such means of killing by wholesale be proper, why should not smaller mines be used?"[40]

Rains stressed the practical nature and tactical effectiveness of his idea. "A shell which can be prepared and unprepared in a moment, and a sentinel to keep our own people off, are all that is wanted for our protection," he argued. If even one torpedo went off and killed an enemy soldier, it would strike terror into the hearts of the victim's comrades. The explosion would offer warning to the Confederates of an enemy approach, serving as better sentinels than flesh-and-blood soldiers. "These shells give us decided advantages over the foe invading our soil, especially in frustrating night surprises, requiring but little powder for great results in checking advancing columns at all times."[41]

Rains's argument appears plausible on the surface, and thus, he began a process of weakening opposition to landmine use in the Confederate army. But his theory has several flaws. First, he focused only on planting mines to check a pursuer or to defend a position. Even Northerners conceded that this fell within the moral purview of established codes of warfare. Rains did not discuss torpedoes as booby traps, which is what really angered Northerners and led them to condemn the torpedo as immoral. In short, as far as the moral realm was concerned, Rains carefully sidestepped the key issue to be discussed.

Secondly, Rains exaggerated the emotional impact of a landmine explosion on the enemy. As we have seen of the Federal reaction to the Yorktown incident, it did frighten, anger, and embitter the survivors, but by no means did it demoralize them. If anything, it made the Federals more determined to fight, if for no other reason than revenge. In the tactical realm, Rains expected far too much of his devices.

Nevertheless, his argument, shaped as it was, slowly began to have an effect. Daniel Harvey Hill, for one, needed no persuasion. When he read Sorrel's May 11 dispatch to Rains, he also endorsed it. "In my opinion all means of destroying our brutal enemies are lawful and proper." He erroneously believed that Rains's May 6 torpedo planting on the Williamsburg Road had completely stopped Federal pursuit.[42]

Another twist to the story appeared on May 12, when Joseph E. Johnston became aware that torpedoes had been planted at Yorktown. He learned of it by reading the *New York Herald*'s publication of McClellan's report. Johnston was concerned and wanted to know if it was true. His assistant adjutant, A. P. Mason, wrote to Hill on May 12, enclosing a clipping from the newspaper. Mason also charged Hill with conducting an inquiry "to ascertain if there is any truth in the statement, to find out if there were any torpedoes placed, and, if so, when, where, and by whom."[43]

In response to Mason's inquiry, Hill wrote to Rains on May 13, and the brigade leader disclaimed any responsibility in the matter. He reminded Hill that his command was among the first to leave Yorktown, "and consequently I know nothing of the location of 'torpedoes' at the places mentioned, nor do I believe it, as wells or springs of water, barrels of flour, carpet-bags, & c., are places incompatible with the invention." Rains ignored the fact that he had planted a booby trap in the Seminole War, while he laid claim to the concept of using artillery shells as landmines. "That invention is strictly mine," he wrote, "for the use of which I have never been called to account."[44]

Rains fully admitted the torpedo planting on the Williamsburg Road on May 6, but he denied placing the booby traps encountered by the Federals at Yorktown. The brigade leader did admit that he had planted some mines at Yorktown. Early in the confrontation, he placed a few torpedoes in front of "a salient angle" of the earthworks "to destroy assailants and prevent *escalade*. Subsequently, with a similar view, they were placed at spots I never saw."[45]

This last sentence clearly shows that Rains was aware of the widespread planting of mines late in the confrontation at Yorktown, but he refused to identify who did it. Most likely, he was responsible. About 1874, when writing a book on torpedo warfare, Rains added another bit of information concerning this issue. He indicated that he wanted to plant torpedoes along a road leading to the earthworks while McClellan was still moving toward Yorktown, but civilians were using the road to flee toward Richmond. He "concluded not to use the subterra shells without due notification."[46]

Rains ended his reply to Hill's letter of inquiry by strongly condemning the Federals for firing artillery at civilian targets. He cited the bombardment of New Bern, North Carolina, his native city, in March 1862 and the bombing of Yorktown a month later. In both places the Unionists offered no warning "to innocent women and children." He called these "fiendish acts unknown among civilized nations, reversing the scriptural text that it is better for ninety-nine guilty persons to escape than for one innocent to suffer."[47]

Thus, Rains justified his developing concept of torpedo warfare on moral as well as tactical grounds. The technological part of it was relatively easy—selecting artillery shells, digging them into the ground, and fixing percussion fuses. He sought to remain within the traditional interpretation that placing such devices in a way to hinder active operations of the enemy was justified. But his moral argument smacked only of revenge and retaliation, neither of which fell into the traditional interpretation. Astonishingly, Rains did not even accept that booby traps had been set at Yorktown, although it was impossible he could never have known they existed.

But everyone among the Confederates took it for granted that the booby traps at Yorktown had been real. Faced with possible censure, Rains now began desperately to seek official approval of his concept of torpedo warfare. He sent a copy of Sorrel's dispatch of May 11 and his endorsement on it to the secretary of war. George W. Randolph, a former lawyer, commanded a clear view of this issue when he added his endorsement to the Sorrel dispatch. It was acceptable to plant devices in or near earthworks to hinder an assault or to place them in rivers and harbors to destroy enemy vessels. "It is not admissible to plant shells merely to destroy life and without other design than that of depriving your enemy of a few men, without materially injuring him." According to the sectary of war, Rains's desire to exact revenge for the shelling of New Bern and Yorktown was clearly out of order.[48]

Randolph offered a suggestion, however, to settle this issue, which helped Rains a great deal. He noted that in a dispute between a superior officer and his junior, the wishes of the former should hold sway. Alternatively, Rains could be taken away from Longstreet's command and placed in charge of defenses along the James River, where the planting of underwater mines was perfectly acceptable. Randolph explained this course of action in a letter written on May 20, and Rains did not hesitate to act on it. "I wrote immediately that I preferred the latter, & was assigned accordingly," Rains recalled. He regretted leaving his brigade but was anxious to get away from Longstreet, "who is a drinking man, & said to be an inebriate." Moreover, Rains had outranked Longstreet in the prewar army and felt he would never receive due credit for anything he accomplished under him.[49]

It is not easy to make full sense of Rains's actions after he was called out concerning the Yorktown torpedoes. He certainly tried to cut a fine point in admitting some things and skirting responsibility for others. Rains admitted to a desire to plant mines on a road that McClellan was using to approach the town but refrained because they might have harmed civilians. He admitted to actually planting devices in front of one salient early in the confrontation but not to planting anything more until May 6, well north of Yorktown, on the Williamsburg Road. His language indicates that he probably knew other devices had been planted, but he refused to admit it or identify who might have been responsible. Most importantly, Rains carefully avoided any hint that he had been involved in booby-trapping the place even though he admitted to a burning motive to do so.

We have Grover's testimony that he personally saw Rains and a man named Gray planting torpedoes. This could well be M. Martin Gray who, at the rank of captain, was in charge of the submarine mines at Charleston during the summer of 1863. Gray also planted numerous landmines fronting Battery Wagner on Morris Island at that time. Admittedly, a deserter-prisoner who is not fully identified and whose testimony is relayed to us secondhand does not provide conclusive proof. Moreover, there are very few personal accounts by men who served in Rains's Brigade dating to the Yorktown operation, and they do not mention anything about torpedoes.[50]

Rains also admitted that on May 6, while arranging for his self-appreciated planting of shells on the Williamsburg Road, he "happened to have" some artillery fuses to put into them. This language implies that he had them in his personal possession. Why the commander of an infantry brigade would carry artillery fuses in his pocket is an interesting question. There was no battery attached to his brigade at this time, so it is safe to assume he carried them in case he could find an opportunity to plant torpedoes.[51]

But the most compelling evidence against Rains comes from a postwar account by John W. Minnich, a Louisiana heavy artilleryman at Yorktown. The detachment he belonged to, led by Lt. William Schirmer, manned a forty-two-pounder seacoast gun in Fort Magruder. They fired their last round at 1:30 a.m. of May 4 and then were led out of the fort by a lieutenant serving on Rains's staff. This man, who might have been a Lieutenant Tyler, Rains's aide-de-camp, told them to walk in single file and to the side of the roadways. He also told the artillerymen "that Rains had caused torpedoes of his 'own invention' to be planted in the road." As soon as the group reached the northern limits of Yorktown, the lieutenant told everyone "[they] were out of any danger from torpedoes." Minnich felt very uncomfortable about this. "I did not at the time, nor have I since, approved of his action in so doing," he wrote of Rains after the war. Minnich was not alone. One of his comrades "vented his opinion on the matter in very plain terms as we emerged from the danger zone: 'This is barbarism,'" he muttered.[52]

Minnich, who wrote several articles about his war experiences, all of which ring true, seems to be a fully reliable witness. The staff officer would have been in a position of authority on the subject of whether Rains was responsible for all the plantings at Yorktown; he stated point-blank that his commander was in charge of the proceedings, even boasting of it. Another Confederate staff officer had no doubt that Rains was responsible for the landmines at Yorktown. James W. Ratchford of Hill's staff thought it was a good idea because he believed the torpedoes demoralized the Federals.[53]

Union losses at Yorktown brought the issue of landmine use to the forefront of public discussion both in the North and South. The episode therefore became a watershed in the history of landmine use in American military history. It enraged thousands of Northern soldiers and civilians, a reaction shared by at least some Southerners as well. The Northern reaction would never change for the rest of the war and would continue even into the postwar era. But among the Confederates, resistance to the use of landmines quickly eroded until it was fully acceptable even to those who initially questioned the morality of their deployment.

Importantly, the Yorktown incident flushed out Gabriel J. Rains from the shadows within which he had previously indulged his interest in subterranean explosives. His Fort King booby trap had not become public knowledge and still remained an obscure fact of his early career for a long while after Yorktown. That Fort King incident demonstrates that Rains had the knowledge and the willingness to set booby traps for his enemy, adding another layer of support to the conclusion that he was responsible for the similar traps set at Yorktown. Both to save himself from censure and to promote a growing conception of the larger issues involved in a torpedo style of warfare, Rains took the offensive after Yorktown. His objective was to vindicate himself, his "invention," and his vision of a new, morally acceptable style of combat.

Chapter 4

Vicksburg, Port Hudson, Jackson

"Legitimate Weapons of Warfare"

The fallout from the Yorktown torpedoes gave Rains his first opportunity to concentrate entirely on landmine warfare, but it took some time for the military authorities to sanction this move. Several weeks elapsed before the order to shift him from infantry command to full-time mining efforts could be issued.

Meanwhile, Rains participated in his first and only Civil War battle when Johnston attacked McClellan's army near Seven Pines, ten miles east of Richmond, on May 31. Hill's Division advanced along the Williamsburg Road with Brig. Gen. Robert E. Rodes's Brigade on the right (or south of the roadway) and Rains's Brigade followed as a second line. The two brigades maneuvered over wet and muddy ground as Rodes fixed Federal attention, while Hill sent Rains to conduct a flanking movement against the Union line. Rains marched his men across the wet ground until they were in position to fire at the Federals. He then stopped, and the troops lay on the sodden earth to open fire. The musketry "struck them like an avalanche," Rains reported. It was "most opportune and important" in Hill's words and played a key role in forcing the Federals to evacuate their first position.[1]

But after that initial success, Rains faltered. At the second Union line, Rains failed to follow through when Hill ordered him to conduct another flanking move. Instead, he remained a short distance behind Rodes's command, extending a bit on the right toward the Federals but falling well short of turning the Union flank. Meanwhile, Rodes pressed forward into devastating musketry. His Sixth Alabama lost 60 percent of its manpower before Rodes ordered the brigade to fall back. He was bitterly disappointed. "I feel decidedly confident that if we had been properly supported," Rodes asserted, the second Federal line could have been taken. Hill agreed, adding that five hundred losses in Rodes's Brigade could have been avoided if Rains had done what he was supposed to do. Later in the day, other Confederate troops managed to flank this second Union line, but McClellan's army steadied itself and held a third position with no difficulty.[2]

In typical Rains fashion, the brigade leader took credit in his report for the capture of the first Union position but did not explain why he failed to help Rodes at the second line. His superiors were keenly aware of that failure, however, and were eager to shift him to torpedo work. Two weeks after the battle, orders were issued to send Rains to the James River, and Longstreet quickly implemented the transfer on June 17. Johnston had been wounded on May 31 and was replaced by Gen. Robert E. Lee, who approved the transfer. After reading Hill's comments on Rains's performance, he admitted that it came as no surprise to him that Rains faltered at a critical moment.[3]

Of course, this was not the way that Rains portrayed his transfer from infantry command to torpedo work. After the war he explained it as a plea from the new army commander to do something about the one hundred Union vessels plying the James River. "It required submarine inventions to checkmate and conquer them," he wrote of the ironclads in that fleet. Although admitting in one sentence that the Russians had placed submarine mines near Kronstadt during the Crimean War, Rains nevertheless bragged in another sentence that the first underwater torpedo he placed in the James River was "the primogenitor and predecessor of all such inventions."[4]

His superiors transferred Rains from the Army of Northern Virginia only in part because of a desire to do something about the Union fleet on the James River. The rest of their motivation was a desire to get an inferior battlefield commander out of the army. He would never lead troops in the field again.

On June 18 Rains was assigned "to the charge of the submarine defenses of the James and Appomattox Rivers." Commanders along both streams along with ordnance and quartermaster officers were instructed to cooperate with him. Despite his postwar self-promotion, Rains was by no means the first man to place underwater mines in the James River. Ever since August 1861, Matthew F. Maury had been working on the problem as head of the Confederate Naval Bureau of Coast, Harbors, and Rivers Defense. He placed fifteen submarine mines close to Chaffin's Bluff by June 19, the day after Rains was officially assigned to duty along the James River. Maury was sent elsewhere when Rains took over the river defense.[5] Rains inherited Maury's handiwork. He reported to his brother that two mines Maury had planted had been torn from their anchors by a spring freshet and were not found. Eleven of the others contained 150 pounds of gunpowder, and two held 2,000 pounds of powder. All were securely anchored in a pattern to bar forward movement of Union vessels, and all were rigged to fire with a Wollaston battery, an electrochemical device. Rains preferred contact fuses in the new mines he deployed that summer along the river.[6]

The fifty-nine-year-old Rains kept in close contact with his younger brother, George Washington Rains. Fourteen years his junior, George had graduated from West Point in 1842 and had served in the Engineers and the Artillery. He had fought in the Mexican War and resigned from the army in 1856 to become an industrialist. George was assigned to gunpowder production when he joined the Confederate army, rising to the rank of colonel by mid-war. He was in charge of the important powder works at Augusta, Georgia, by the time Gabriel wrote to him in July 1862. "I have not heard from my family in Marshall, Texas since 27 April last," he complained. His family consisted of a wife and seven children. But Rains spent more time complaining to his brother about his career than about his domestic life. "My position has been somewhat mortifying in being put under men, my juniors whose military requirements were not superior to my own." Overlooking his failure during the second phase of the battle at Seven Pines, Rains expressed a degree of pride in his performance in that engagement. The best news, in his view, was Lee's victory over McClellan in the Seven Days campaign of late June and early July.[7]

Lee's success in the Seven Days greatly reduced the necessity for mining the James River. As a result, Rains was ordered to Wilmington, North Carolina, on September 3, 1862. Placed in command of the district of Cape Fear, he was expected to plant submarine mines to protect the seaport of Wilmington, which was located several miles up the river. His progress was delayed by the onset of yellow fever in the area, which inhibited the activity of his staff and crew. Rains also had to send men and material to other port cities from time to time.[8]

In November 1862, just when he started laying a number of mines in the Cape Fear River, Rains was replaced as commander of the district. This turn of events angered the frustrated general. He wrote a long letter to Jefferson Davis on November 24, laying out the details of his military career and expressing the belief that he was the victim of discrimination because of the "Torpedo Shells" that he had developed. Davis passed the letter on to Adjutant General Samuel Cooper for his opinion.[9]

This letter had its effect whether because Davis believed in using any available weapon or because he recalled Rains's participation on the court martial that exonerated him nearly thirty years before. The Confederate Congress had passed an act in October 1862 authorizing the creation of a Naval Submarine Battery Service for underwater mines and a Torpedo Bureau, the latter presumably for landmines. But the act was not presented to Davis for his signature until near the end of the war. Despite this, Davis moved forward to make both agencies a reality. He called Rains to Richmond early in December 1862 and placed him in charge of the Conscription Bureau because the Torpedo Bureau was not yet an official agency. Rains worked it out so that his acting

Figure 4.1. Gabriel J. Rains. A prewar portrait of the man most responsible for introducing the contact mine to the Civil War and to global history. Rains's most important technical contribution was developing a sensitive detonator. He also was the first to develop a doctrine for the deployment and use of landmines, although that doctrine was not publicized during his lifetime. An inveterate self-promoter, he greatly exaggerated the tactical benefits of his device. (LC-DIG-cwpb-07529)

adjutant general, Lt. Col. George W. Lay, quietly handled most of the business of the Conscription Bureau to give Rains time to work on torpedoes.[10]

John B. Jones, head of the Confederate War Bureau, was temporarily assigned to help in the creation of the Conscription Bureau. "I like Gen. Rains," he wrote in his diary on January 8, 1863. Rains came to visit Jones every day and led him to believe he was "engaged in some experiments to increase the efficiency of small arms." Rains also promoted his war record by telling Jones his brigade "retrieved the fortunes of the day" at Seven Pines, criticizing Johnston and Randolph for "giving up of the Peninsula, Norfolk, etc." By January 24 Jones had a more accurate view of what Rains was doing while pretending to head the Conscription Bureau. The general was "making a certain sort of primer." One of them went off in his hand that morning, "injuring his thumb and finger. He was scarcely able to sign his name to official documents to-day."[11]

The next day, January 25, Rains was more open with Jones in his conversation. He discussed the sensitive primer and assured him "he would not use such a weapon in ordinary warfare; but has no scruples in resorting to any means of defense against an army of Abolitionists, invading our country for the purpose, avowed, of extermination." Rains claimed great success in stopping "the army of invaders" along the Williamsburg Road on May 6.[12]

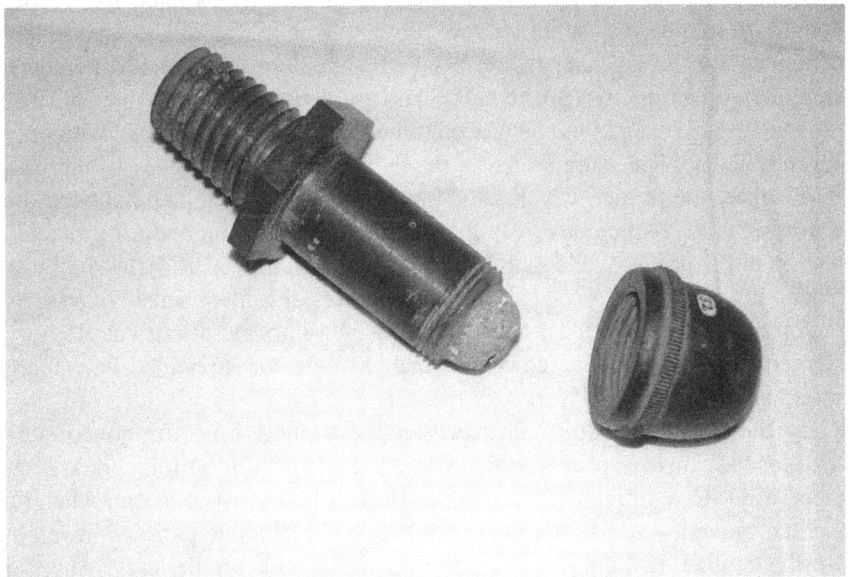

Figure 4.2. Rains's Sensitive Fuse, Ten-Inch Mortar. This is an example of Rains's sensitive fuse, designed for a ten-inch mortar shell. Note the protective cap to prevent accidental detonation. (AHC and Herbert M. Schiller, *Confederate Torpedoes*)

This sensitive primer was the only technological contribution Rains made to the development of torpedo warfare. All his other contributions fell into the realm of tactical innovations. Exactly when he started to work on improving the percussion fuse to make it more sensitive to pressure is unclear. Brig. Gen. William F. Barry, McClellan's artillery chief on the Peninsula, described the fuses found with the Yorktown torpedoes as the "ordinary artillery friction primer." In the letter to his brother dated July 5, 1862, Rains referred to "one of my sensitive priming tubes covered over with thin copper [and] soldered watertight." These he planned to apply to the James River underwater mines. After the war, as we have seen, he referred to having sensitive primers with him on May 6, when he planted artillery shells on the Williamsburg Road. Yet Jones clearly states that he had "invented a new primer for shell, which will explode from the slightest pressure," and Jones knew it was to be buried in the ground as an antipersonnel weapon. It is possible that Rains had developed the sensitive primer early in the war and used it at Yorktown and on the Williamsburg Road. He might have been trying to improve it in January 1863, when one went off and injured his hand. We cannot tell exactly when he developed it due to Rains's habit of ignoring crucial elements of truth and spreading exaggerations about his work.[13]

Rains wrote that his creation should be transported "in boxes packed with cotton" so they would not be damaged in transit. A minute examination of the sensitive primer is possible because, a hundred years later, five torpedoes were found on a battlefield, probably at Mobile, by relic hunters. They contained the Rains fuse. It consisted of potassium chlorate (50 percent), sulphuret of antimony (30 percent), and pulverized glass (20 percent). A protective seal made of copper covered the top. It could be crushed with only seven pounds of pressure so that a man's foot or a horse's hoof would easily detonate the primer, which set off a short, quick match that exploded the powder charge. A protective cap that could be screwed on and off protected the top of the primer while in transit and could be removed after the shell was placed in the ground. A few additional examples of Rains's sensitive primers are on display at the US Military Academy Museum and the Atlanta History Center. They were discovered at Savannah, Mobile, the Richmond-Petersburg lines, and Port Hudson.[14]

By the spring of 1863, more because of support from the government than for his sensitive primer, Rains was the go-to person for torpedoes. Brig. Gen. W. H. C. Whiting advised Daniel Harvey Hill to write to him when he wanted to deploy sub-terra mines at fortified Confederate garrison towns in North Carolina. Hill did so and Rains told him that his brother at the Augusta Arsenal could supply "whatever sensitive priming-tubes you want. . . . I went there and instructed in their make, and they have been furnished Mobile, Savannah, Charleston, &c., by the thousand." As for the explosive device, it consisted of artillery shells that Hill could easily obtain. Rains also gave instructions to Maj. J. L. Cross, who later served as assistant adjutant general on Hill's staff, "in the use of both subterra shells and submarine mortar shells." He was spreading technical information to promote torpedo warfare as widely as possible, in addition to flooding the Confederacy with his sensitive primer.[15]

In Rains's mind, self-promotion always went hand in hand with promoting torpedo warfare. To cap that campaign, he drafted a book by the end of May 1863, included drawings in it, and presented the manuscript to Davis for his perusal. A copy of this manuscript has not survived, but it most likely was similar to the "Torpedo Book" that Rains wrote in or about 1874. "It caused him to enter with zest into the schemes," Rains told Hill about Davis's reaction to his book on May 30. The result was that the Confederate president wanted Rains to go into the field personally to apply his concepts.[16]

But Davis objected to Rains's suggestion that the book be published. It would be impossible to keep the concepts secret, no matter how few copies were printed or how carefully the Confederates tried to keep it under cover. Instead, Davis assumed Rains had sent him a duplicate copy, not the original, and wanted to use extracts from it to instruct various commanders in the field about how they could proceed with the use of torpedoes. He was willing to give the manuscript back to Rains, but there is no evidence that this was done or that Davis followed through with his idea to extract information from it for the use of his generals.[17]

VICKSBURG

The appearance of Rains's book coincided with a period of intense anxiety concerning the fate of Vicksburg. Maj. Gen. Ulysses S. Grant had finally turned the corner in his long campaign for the strategic Mississippi River town in May, approaching it from the south and east in a move to cut the place off from the rest of the Confederacy. Grant separated the forces commanded by Lt. Gen. John C. Pemberton and a much smaller force led by Joseph E. Johnston, driving the former into the city, while leaving the latter at the state capital of Jackson. This was the situation when Davis ordered Rains to report to Johnston "for duty in connection with torpedoes and subterranean shells." The order was dated May 25, but Davis had already written to Johnston the day before that he thought Rains could provide valuable service in the field.[18]

For the only time in his career, Rains balked at an opportunity to use his favorite weapon. He remembered that Johnston had started an inquiry into the use of sub-terra shells at Yorktown and protested to Davis about the assignment, "but it was no go." Rains assumed that Johnston did not approve of torpedo warfare, but that may not have been true. Actually, Johnston merely began an inquiry to ascertain if Northern newspaper reports about the Yorktown torpedoes were true and, if so, to find out who planted them. There is no evidence that anything came of the inquiry and no direct evidence that Johnston necessarily disapproved of their use. He could have been angry that it was done without his approval rather than that it was done at all.[19]

Nevertheless, Rains operated under the assumption that Johnston was his enemy, and he tried to get out of the assignment. Davis refused to allow it. He told Confederate Secretary of War James A. Seddon to write to Johnston on May 27 and avert any friction that may arise from the general's association with Rains. "The President has confidence in his inventions, and is desirous that they should be employed both on land and river," Seddon assured Johnston. The idea was to help Rains sneak through Union lines into Vicksburg if possible. "All reasonable facilities and aid in his supply of men or material for the fair trial of his torpedoes and shells are requested on your part," Seddon continued. "Such means of offense against the enemy are approved and recognized by the Department as legitimate weapons of warfare."[20]

Seddon's letter must have excited Rains, for it represented the culmination of his campaign to promote torpedo warfare. He gained the approval of Davis and, with it, the legitimation of mines on both land and water. For the first time in American history, an official policy of torpedo warfare had been approved and resources committed to make it happen. Rains had access to the letter; he sent a copy of it to Daniel Harvey Hill, who had been a convert to torpedo warfare ever since Yorktown.[21]

Then Rains wrote to Davis on May 31 and almost wrecked his relationship with his most important patron. Again, despite Seddon's letter, he expressed doubts about whether he could work effectively with Johnston. Davis's patience was greatly tried. In fact, he was surprised that Rains was still at Richmond. Davis dismissed both reasons for Rains's hesitancy in his reply of June 3: first, that he could not reach the theater of operations in time to be of service and second, that Johnston would not cooperate with him. If Rains did not want to go, Davis would send someone else because he was desperate to employ all means to save Vicksburg.[22]

In fact, there already were Confederate torpedo operatives in the Mississippi Valley. They had used electrically detonated submarine mines to sink the USS *Cairo* in the lower reaches of the Yazoo River north of Vicksburg in December 1862. In another example of his overpowering ego, Rains had assumed that one of his sensitive primers had brought the vessel down, but in reality, his primer had nothing to do with it. Davis, however, assumed Rains was right and told him that the sinking of the *Cairo*, more than any other factor, convinced him that Rains was a genius with a new mode of warfare that the Confederacy needed to employ.[23]

Now that he had painted himself into a corner, Rains was compelled to go to Mississippi. Sometime between June 3 and June 29, he traveled to Jackson, the base of Confederate efforts to relieve the siege of Vicksburg, and set up shop. He paid $200 for a room wherein he set up a laboratory "for making sensitive primers" and spent $165.50 to transport "material & c for Subterra Shells & Sub Marine mortar Batteries" by rail from Richmond and Augusta to Jackson.[24]

But Rains was deeply dissatisfied with his new assignment. Writing to Gov. Zebulon B. Vance in North Carolina on June 29, he bragged that Davis wanted him in Mississippi so badly as to declare that "my services at Vicksburg [were] equivalent to 5000 men." But the Union grip on Vicksburg was too tight to allow him in, and Johnston concentrated on building up his forces rather than trying to raise the siege. There seemed to be nothing for him to do, and anything that was possible in the way of torpedo warfare could be conducted by "instructed officers." Rains felt he was wasting his time at Jackson.[25]

Instead, he keenly wanted to be assigned to North Carolina. Rains told Vance that if he had been in command of its defense, the Federals "would never have had a foothold" in the eastern portion of his native state. Rains felt that his prior experience in the US Army justified higher rank in the Confederate army. "As a son of N. Ca. I have been galled by the acts of both friends and foes," he complained and then suggested that Vance appoint him lieutenant general of state forces. Of course, Vance ignored this suggestion.[26]

While Rains fumed in Jackson, Confederate engineers inside Vicksburg created a few explosive devices to aid their defense. Engineer Capt. Powhatan Robinson planted a fougasse at a traverse along the sector held by Brig. Gen. Stephen D. Lee's Brigade late in the siege. Earlier, Capt. James T. Hogane, another engineer officer, had placed obstacles along River Road that ran between Fort Hill and a series of water batteries on the northern sector of the defense perimeter. These obstacles included "three deep ditches across the road," sharpened stakes in the bottom, and "mines" between the ditches. None of these devices were exploded during the siege.[27]

PORT HUDSON

Simultaneous with the siege of Vicksburg, Federal forces under Maj. Gen. Nathaniel P. Banks besieged a small Confederate garrison at Port Hudson, Louisiana. Lt. Frederick Y. Dabney, chief engineer of the defenses, was alarmed when Federal sappers approached the sector held by the First Mississippi and Fifteenth Arkansas, which included Battery No. Eleven. He authorized the planting of "a number of 8, 10, and 13-inch shells" in "the scarp wall of the ditches as torpedoes." Capt. L. J. Girard of the ordnance department arranged the shells and planted them. He then used wires to connect each one with the interior of the work. The intent was to pull the wires and set off friction primers in the planted shells. A Federal who dug one up discovered "a five-gallon demijohn, filled with ounce bullets, with a bursting charge in the center." It was rigged with "a common percussion primer," like the ones artillerymen used to explode shells. As at Vicksburg, these improvised landmines did not employ Rains's sensitive primer, which was set off by contact rather than by a lanyard. Also as at Vicksburg, none of these devices were exploded during the siege.[28]

As demonstrated by the deployment of mines at Vicksburg and Port Hudson, Rains was not the only man interested in the use of explosive devices to aid the tactical defense. He had by the summer of 1863 achieved far more prominence than anyone in the developing concepts of torpedo warfare, but he was not wholly responsible for it either. In neither siege did Confederate engineers use his sensitive primer, and no one could claim proprietary rights in the idea of using artillery shells as sub-terra mines. Whether Rains could have achieved more than the engineers at Vicksburg or Port Hudson if he had managed to get into either place is an open question.

JACKSON

After the fall of Vicksburg on July 4, Rains had an opportunity to employ his devices when Maj. Gen. William T. Sherman led a large force eastward to deal with Johnston's army. "Genl Rains should now fully apply his invention," Davis told Johnston, who waited until he was ready to evacuate Jackson on the night of July 16 to do so. Rains proudly reported in his postwar "Torpedo Book" that Johnston came to seek his help, explaining his plan of evacuation and asking "If I could not check the advancing enemy . . . by the shell." They agreed to plant some sub-terra mines on the west side of the bridge over Pearl River, immediately east of Jackson. Rains enlisted the help of Lt. Col. James P. Parker of the First Mississippi Light Artillery to select the biggest artillery shells available but then changed his mind about the placement. Rains and Parker planted them at the east end of the bridge and on two roads that led east from that point. Parker later decided to plant some at the west end of the bridge, too, boxing the main crossing of the river with landmines.[29]

Unlike the Yorktown episode, Rains admitted responsibility for the torpedoes planted at Jackson. Lt. H. F. Scaife of the Macbeth Light Artillery recalled hearing Rains "imagining one day the terrible fate that awaited the Federals, when they passed over these explosives." As Scaife's unit crossed the river, "guards were stationed at different places where he had buried some of these weapons, to turn our troops out of the road so as to prevent the destruction of our men." Fred Joyce of the Fourth Kentucky, C. S., recalled "moving cautiously along a sand road" during the nighttime retreat, hearing repeated warnings to avoid "the dangerous torpedoes" in it.[30]

When the Federals entered Jackson and explored its surroundings, they encountered these devices. Several men were injured, but the exact number is hard to pin down. Sherman only reported that a citizen and two soldiers were wounded by torpedoes, but Hugh Boyd Ewing noted in his diary that six men of his Fifteenth Corps brigade were injured by them.[31]

After the war Rains claimed that he was concerned civilians would get entangled in the landmines. According to Union accounts, that not only happened but may well have hurt as many citizens as Federal soldiers. The historian of the Forty-sixth Indiana testified that a horse drawing a cart that carried a man, two women, and two children stepped on a torpedo. "The horse, cart and people were distributed over the road."[32]

Confederate prisoners also were hurt by these devices. A newspaper correspondent reported that a Federal cavalryman escorting a Rebel prisoner trod on a torpedo several hundred yards from the river by one of the roads that Rains and Parker had made dangerous. "The horse was literally split wide open by the explosion, and the rider almost instantly killed." The prisoner was hit by a shell fragment below the thigh, which shattered the bone, and he was not expected to survive.[33]

Capt. Jacob Roemer of Battery L, Second New York Light Artillery had no idea of danger as he rode toward the west end of the Pearl River Bridge, until a sentinel sprang up to grab his bridle and stop him. "For God's sake, Captain, don't go another step or you are a dead man," he shouted. The two talked for a while until a Rebel prisoner who was nearby began to brag about how the Confederates had fooled Sherman. His conversation was cut brutally short when the prisoner stepped on one of Parker's torpedoes. It exploded, taking off both legs, and the man quickly bled to death. "While sitting there looking at that dead rebel, I could not but wonder at the quantity of blood in a human body," mused Roemer. "I have seen hundreds of wounded men, but I have never witnessed another scene like that. I felt very sorry for this poor fellow, who had lost his life through the actions of his own people. It was a most sickening sight."[34]

The danger did not end that day. On the next, July 18, Curtis P. Lacey of the Fifty-fifth Illinois saw two torpedoes explode, but they did not hurt anyone. He probably witnessed a deliberate detonation as the Federals were busy demining after Johnston's evacuation of Jackson. They used Confederate prisoners to dig up the mines.[35]

Sherman's command held Jackson from July 17 to 23, and it seemed to Capt. Henry S. Nourse of the Fifty-fifth Illinois that nearly every night, a fire or two broke out in town. He was sure it was done in retaliation for the torpedoes and for the Confederate practice of poisoning wells in the area. Other Federals agreed that the landmines at Jackson embittered their comrades. These "cowardly contrivances" hurt civilians and soldiers alike.[36]

Rains gloried in the destruction wrought by his sub-terra shells at Jackson, relying on hearsay to gauge their tactical effectiveness. He believed reports that when pursuing Union cavalry came across them on the road toward Brandon, Mississippi, they were so frightened that all pursuit was called off. One horse was thrown "bodily in the top of a tree," he asserted. For the second time (following the Williamsburg Road incident of May 6, 1862), landmines had saved Johnston's retreating command, in his view.[37]

All of this was nonsense based on Rains's obsessive desire to grab all the attention he could get. A mixed column of Federal troops crossed Pearl River on pontoon bridges and pursued Johnston all the way to Brandon, thirteen miles east of Jackson, but it encountered no torpedoes anywhere along the way. Then Sherman called off further pursuit because he intended to abandon Jackson and return to camps at the Big Black River, closer to Vicksburg. There were "no good military reasons for a longer stay at Jackson," as Sherman put it, or for a further push east of it. The Federals called off their pursuit because of their larger strategic goals rather than because of Rains's sub-terra shells.[38]

Yet Rains and his supporters believed what they wanted to believe. "My old friend, Gen. Rains . . . stopped and stampeded Grant's army" with his torpedoes, wrote John B. Jones in his diary. Hundreds were killed, and the rest fled back to Vicksburg. "This invention may become a terror to all invading," Jones concluded. He was a bureaucrat planted in Richmond, dependent on rumor and Rains's own fractured view of reality for his information. But Pvt. Welburn J. Andrews of the Twenty-third South Carolina, a member of Johnston's command, also believed the rumors. Davis had no difficulty embracing what he hoped was the truth about the effectiveness of Rains's shells at Jackson. After the war he asserted that Rains had the best opportunity to have an effect on the course of a campaign during Johnston's retreat from Jackson.[39]

A balanced assessment of the torpedoes that Rains and Parker placed at the crossing of the Pearl River does not support this overblown Confederate view of their effectiveness. From a technical viewpoint, they worked very well. But from a tactical perspective, they became nothing more than an annoyance to the Federals, an annoyance that angered them yet failed to have any effect on their movements. As at Yorktown, Union commanders assigned guards to warn others of the danger and forced Confederate prisoners to find and remove them. Rains and Parker quickly and easily planted the torpedoes at Jackson, but the Federals just as quickly and easily dealt with them.

The level of loss was minimal as well. Probably no more than half a dozen Federal soldiers were injured or killed, according to the admittedly sketchy evidence. Significantly, for the first time in the war, Southern civilians were killed and injured by torpedoes. In fact, there seems to have been about as many citizen victims of landmines at Jackson as military victims, and at least two of those military victims were Confederate prisoners. Victim-activated landmines could not distinguish between friend and foe. They were just as much of a menace to innocent civilians as to enemy soldiers.

Chapter 5

Fort Wagner, Fort Esperanza
"Scarcely Civilized Warfare"

Immediately following the limited use of landmines at Jackson, Confederate deployment of torpedoes picked up considerably in the summer of 1863. Union offensives against two coastal islands, one in South Carolina and the other in Texas, led Confederates to employ landmines to supplement their fortifications. While the Texas example introduced nothing new in the history of landmines and resulted in no detonation of the devices, the South Carolina example witnessed the planting of the first true minefield in American history, the first sapping operation through a minefield in world history, and the killing and injuring of a number of Federals and a handful of Confederates alike.

FORT WAGNER

A Federal drive to capture Morris Island, located on the southern flank of the entrance to Charleston Harbor, led the Confederates to employ every means at their disposal to hold on to the strategically placed island. Brig. Gen. Quincy A. Gillmore, a thorough engineer, was appointed commander of the Department of the South on June 12, with the assignment of reducing Rebel forts on the island and bombarding Fort Sumter as a prelude to a major assault aimed at capturing the port city.[1]

With eleven thousand infantrymen at his disposal, Gillmore landed on the southern end of Morris Island on July 10 and advanced until within range of Fort Wagner, the major Confederate work on the island. Constructed as a battery that mounted artillery in the summer of 1862, Gen. P. G. T. Beauregard had ordered it improved to include infantry positions in the fall of that year, upgrading its designation from a battery to a fort. The sea face stretched 210 feet and had three gun emplacements, while its land face was 600 feet long and contained five gun emplacements. Two more artillery platforms were constructed behind an extension of the land face that went to the high-tide mark, and a deep ditch fronted the land face. Parapets were up to fourteen feet thick at the top and nearly fifteen feet tall. Morris Island consisted mostly of shifting sand that had accumulated at the edge of the coastal marsh. While the southern end was one thousand yards wide, it narrowed a great deal until only twenty-five yards of sand capable of supporting men existed just south of the fort.[2]

These terrain difficulties constrained Gillmore's choices when he approached Fort Wagner. He tried a quick attack with three regiments on July 11 that failed. Then, a week later, he tried a second attack with two brigades on July 18. Far bigger than the previous attempt, this bloody assault also failed, although some Federal troops managed to occupy the corner of the fort for a time. Having lost 1,515 out of 5,000 men in the second attack, Gillmore settled down to siege operations.[3]

"The contest, therefore, is now purely one of military engineering," Beauregard told Confederate Secretary of War James A. Seddon on July 20. He ordered the construction of another work, Battery Gregg, to protect the wharf at the north end of Morris Island, the only line of communication linking Wagner's 1,200 men with Charleston. Beauregard did not hesitate to use landmines as an adjunct to the fort's defense. "Order large numbers of Rains torpedoes established in advance of Battery Wagner at proper time," he told a subordinate on July 10. Beauregard made it clear that Fort Wagner "must be held and fought to the last extremity consonant with legitimate war." He also wanted Rains torpedoes to be planted in the roads and approaches to the Confederate line on James Island. This large island, located just west of Morris Island, offered Gillmore his most direct line of advance toward Charleston.[4]

On July 11, the day of the first Federal attack, Capt. M. M. Gray placed fifty-seven artillery shells with Rains's sensitive primer in a group five to twenty yards in front of the land face of Fort Wagner. Gray probably had been the officer who helped Rains plant the landmines at Yorktown more than a year before, and now he worked as superintendent of Submarine Batteries at Charleston. Gray waited another week and then conducted a careful examination of these plantings right after the Federal attack of July 18. He found that twenty of the fifty-seven had exploded, "many of them by the enemy's shells; others I found damaged by the heavy rains previous to that date." He made no conclusion about whether any had been exploded by attacking Union infantrymen, but it was possible.[5]

Based on his examination, Gray made more torpedoes but added a waterproof cap to protect the priming mechanism and chose much larger projectiles. He managed to find three fifteen-inch shells with twelve pounds of gunpowder in each and forty-four shells ranging in gunpowder quantities from thirty-three to forty pounds each. On the night of July 20, Gray planted these forty-seven shells from 20 to 240 yards in front of the land face of Fort Wagner. He restricted himself to the ground lying between the fort and a small sand ridge 240 yards in front of Wagner. Each of these devices had a Rains sensitive primer. After the war Rains claimed that Gray's fifteen-inch shells were those fired by Federal naval vessels that had failed to explode. If so, the Confederates were turning them to their own purpose.[6]

Beauregard himself, an accomplished engineer, weighed in with advice about how to plant the torpedoes from his office in Charleston. He warned against placing them too close to Fort Wagner, where Federal artillery fire on the work could fall short and either explode them or throw enough sand about to cover the primer. He advised that some be placed in the bottom of the ditch and others, one hundred yards in front of the work. There is no evidence that Gray did this; Beauregard's advice was written on July 25, five days after Gray had completed the placement of 104 landmines.[7]

Gray and his subordinates made more torpedoes at their workshop in Charleston. By August 12 he had thirty-nine landmines and fifty-eight underwater mines ready for deployment. His men were in the process of arranging fifty more, but he did not specify if they were meant for land or water. In addition, forty-four "submarine shells of Rains' pattern" would be ready by August 20.[8]

The Confederates used a mix of Rains's devices plus improvised mines of their own at Charleston. Rains himself would have an opportunity of observing their handiwork, for on August 20, he was ordered to Charleston. After arrival, Beauregard assigned him to take charge of preparing and planting underwater mines to protect the harbor and the water approaches to the city on September 2. Gray had already placed a number of submarine mines at various locations during the month of August, and it is unclear exactly what Rains did to supplement his work while he was at Charleston.[9]

At least some of these underwater mines were ripped from their moorings by storms and washed up on the beaches, to pose a danger. Justus Schubert, a Prussian military observer with the Confederates, reported that they appeared on the southern side of Charleston Harbor between Fort Johnson on James Island and Fort Wagner on Morris Island. He indicated that the Confederates disarmed those that washed up on James Island and implied that they reused those that appeared near Fort Wagner, planting them as landmines. These underwater torpedoes continued to wash up for months after the Morris Island campaign ended.[10]

Gillmore's troops conducted the "first deliberate breach of a minefield" in world history when they began to sap forward to create parallels across the sand fronting Fort Wagner.[11] It was not the first infantry attack across a minefield. As we have seen, the British did that when they hit the Russian position near the Quarries at Sebastopol in 1854. But the Federals conducted the first sapping operation through a minefield and had to learn how to do it so as to minimize losses.

Federal sappers would begin their introduction to mine warfare only after reaching the sand ridge located 240 yards in front of Fort Wagner. The Twenty-fourth Massachusetts captured that small ridge on August 26, taking seventy Confederate pickets for the loss of only eight men. Those prisoners warned the Federals that they would encounter "formidable torpedoes" beyond the sand ridge all the way to the glacis fronting the ditch along the fort's land face. In fact, they admitted that they did not retreat when the Massachusetts men assaulted, in part, because they were afraid of running through the minefield under fire.[12]

"We had now reached the point where the really formidable defensive arrangements of the enemy commenced," reported Gillmore. He confronted "an elaborate and ingenuous system of torpedo mines." The only way to deal with this obstacle was to dig a sap from the sand ridge on which Gillmore would establish his fifth parallel, instructing the sappers to be careful when they came to a mine.[13]

The Federals hit their first torpedoes on the night of August 26 after several hours of digging forward. They began a sap from the sand ridge, rapidly digging through soft sand under cover of night until reaching a point about one hundred yards from Wagner and near the beach. Then they began to dig toward the land face and here encountered a torpedo that killed a member of the fatigue detail, a corporal of the Third US Colored Infantry. The blast blew his body twenty-five yards away, with his arm resting on the plunger of another torpedo. He was killed instantly, but all his clothes were blown or burned off. Maj. Thomas B. Brooks, aide-de-camp and assistant engineer to Gillmore, was standing twenty yards from the corporal at the time and assumed he had been killed by a shell from Fort Wagner. Only when another torpedo was found later that night did he realize what had really happened.[14]

The next morning Federal observers were shocked to see the corporal's nude body laid out in the minefield. The sight gave rise to bizarre rumors that the Confederates had deliberately tied his body to the unexploded mine as a decoy to lure unwary Federals into danger. An inflammatory woodcut illustration of the incident was printed in *Harper's Weekly* on September 19, 1863, with a clear image of the Confederate keg torpedo under the nude body. In this case, the Confederates were unjustly stigmatized as barbarians for cruelly using the dead African American, but the effect on Northern notions of fair play must have been significant, for the woodcut image was an unusual illustration for its day.[15]

As the Federals continued to work on August 27, the day that many soldiers gaped at the bizarre sight of the corporal's body, they found eight more

Figure 5.1. Death of a Corporal, Third US Colored Infantry. He was the first victim of a landmine during the siege approach to Fort Wagner. The torpedo blew his body twenty-five yards away so that his arm rested on another mine, and it blew or burned off all his clothes. It appeared to some observers as if the Confederates had deliberately tied his body to the unexploded mine to lure more Federals into the mine field. *Harper's Weekly* took full advantage of this incident to print an inflammatory illustration, which also depicts how the Confederate keg torpedo was planted. (*Harper's Weekly*, September 19, 1863)

torpedoes. While trying to move one out of the way by laying a rope around the upper part and gently pulling at it, the device exploded. The Confederates listened attentively to all that was going on in the opposing line and counted three rather than one torpedo explosion on the night of August 26 and another one on the night of August 27.[16]

But Brooks and his men quickly learned how to deal with these torpedoes. They carefully dug around them and lifted the devices out of place without touching the sensitive primer. Lt. James S. Baldwin of the First New York Engineers suggested drilling a small hole through the side of those torpedoes that were fashioned from wood. After making the hole with an auger, he would pour water in to render the gunpowder harmless. This was done with at least some of the torpedoes but was time-consuming. Brooks also experimented with rifle fire, ordering sharpshooters to explode the torpedoes that were visible in front of the Union sap on August 29. It did not work. "Hitting the plunger did not explode them," probably because the plunger responded to pressure from directly above rather than by force exerted at an angle. Some Federal solders amused themselves by throwing things onto the plunger whenever a torpedo was in range of their pitching arms. George Benson Fox of the Seventy-fifth Ohio claimed that at least four devices were touched off this way.[17]

As the Federals dug closer to Fort Wagner, the torpedo danger increased. On the night of August 31, Brooks lost some men when they crept ahead of the sap to protect the diggers from a possible sally by the Confederates. One man touched off a plunger in taking position, and the resulting explosion made casualties of him and two comrades.[18]

Engineer Lt. Peter S. Michie temporarily relieved Brooks of his charge of the sapping operation on September 1. The next day, Michie reported that the sap roller, a large and heavy device rolled forward to protect the sappers, could no longer be used because of the danger of touching off torpedoes. The Federals continued to dig forward without the roller until September 5, when Michie ordered them to stop for a while due to heavy artillery fire at short range. "One torpedo was removed, and four others can be seen near the foot of the slope of the" glacis, the raised ground leading up to the outer edge of the ditch along the land face. Continuing to dig forward on September 6 under cover of naval gunfire, Michie's men reached a point twelve yards from the beach end of the land face late that night. They had removed six or seven torpedoes during this day's digging without accidents. But another landmine killed one man and wounded three others that day.[19]

Brooks was back on duty by this time and noted that because the sappers had veered as close as possible toward the beach, they avoided the worst of the Confederate minefield. By glancing to the left, they could see numerous plungers barely sticking out of the sand covering most of the land front. Daring individuals crept forward under cover of darkness to take a look at the ditch and the parapet. They still had to navigate around a few torpedoes along the way but did so "with a skill which the most bitter experience only could have conferred," as stated in Gillmore's words.[20]

Obviously, the torpedoes which constituted the first true minefield in American military history failed to keep the besieger away. In fact, they hardly delayed Federal sapping operations. The Unionists learned quickly how to deal with them after a couple of ineffective experiments. There was no substitute for careful digging and handling without activating the sensitive primer.

Figure 5.2. Confederate Keg Torpedo. This was one type of landmine deployed by Confederate operatives in front of Fort Wagner during the late summer of 1863. It was originally designed as an underwater torpedo but was used in small numbers to supplement the artillery shells that were more generally used at Fort Wagner. (USMA and Herbert M. Schiller, *Confederate Torpedoes*)

The Federals dug up more than sixty torpedoes during their approach to Fort Wagner, giving them ample opportunity for study. Brooks divided them into three categories according to their size and origin. In the first category, he placed about twenty twenty-four-pounder artillery shells with wooden plugs in the fuse hole and a small hole drilled through the plug. The fuse was fixed in this hole with an explosive mix in a ball at the outer end. "Over all, enveloping the shell, was a cylindrical box of thin tin, painted black. The bottom of this box rested on the cap." The device was buried so that the ball with the explosive mix rested just below the surface and a "slight pressure" was enough to set it off.[21]

In the second category, Brooks placed thirty-seven wooden kegs of ten-gallon capacity, "the ends of which were extended by conical additions." They were similar to submarine mines, and Brooks assumed they had originally been intended for placement in the harbor but were diverted to land use. A hollow plug with a plunger in it to activate a paper tube containing explosive material made up the detonation device. In most of these torpedoes, Gray and his team had placed a board laid on the ground and that rested on the plunger to increase the range at which a Federal would detonate the explosive. Brooks found that in a handful of devices, the board was replaced by "a cap having three arms of iron." Disturbing any one of those arms set off the explosive. Because they were made of wood, the Federals found it easier to dispose of these kegs than the other two varieties of landmines. Using Lt.

EXPLOSIVE ARRANGEMENT.

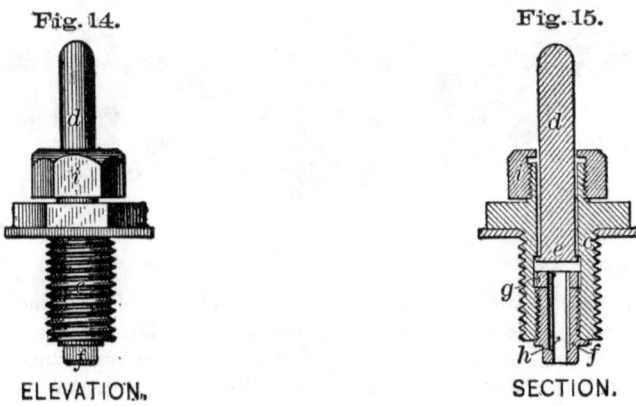

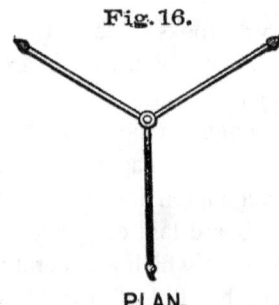

Figure 5.3. Fort Wagner, Fuses and Detonation Devices. This illustration by the Federals shows the fuses and detonation devices used at Fort Wagner. The three-pronged wire expanded the area of danger for unwary soldiers. Illustrations such as this were part of the Union effort to understand and publicize information about these new devices of destruction. (*OR*, vol. 28, pt. 1, 311)

Baldwin's idea, they carefully drilled holes and poured water into more than thirty of them.[22]

Surgeon Samuel W. Gross took the time to examine these keg torpedoes after they were deactivated. He found them to be lager beer kegs, twenty-three inches long. With the conical projections at both ends, they extended to fifty-seven inches in length. The kegs measured fifty-four inches in circumference at the middle, and their iron hoops were covered with tar to waterproof them. Engineers told Gross that a pressure of only four pounds was enough to touch off sixty pounds of powder in each keg.[23]

The third category consisted of three fifteen-inch naval shells planted the same way as the smaller shells. They were fitted with "the metallic explosive apparatus like the wooden ones," Brooks reported.[24]

The torpedoes at Fort Wagner led to losses even after they were taken out of the field and stored at engineer depots. Brooks noted that about a dozen Federals were killed or injured in this way. William L. Hyde of the 112th New York described one such accident that took place the day before the Confederates evacuated Fort Wagner. An engineer was "working over a torpedo, handling it carelessly," when the thing went off. It decapitated the engineer and wounded two African Americans "who were assisting him."[25]

A handful of Confederates became victims of the landmines at Fort Wagner. When Gray planted the devices, one of his men, Thomas McNall, was badly injured after he accidentally crawled upon one and touched it off. A few days later, Beauregard sent a dispatch to Gillmore by flag of truce on the afternoon of July 27. During the ceasefire, two men of Lucas's Artillery Battalion, commanded by Capt. Robert Pringle, wandered into the minefield against advice. "They had been warned, and were called to and ordered to come in," reported Capt. C. E. Chichester, chief of artillery at the fort. On the way back, one of them touched off a mine, and both men were killed. "They demonstrated that torpedoes will explode, when our own men step on them."[26]

The minefield at Fort Wagner was a new development in warfare, and Gillmore and Brooks were interested in its tactical context. They had wondered why the Confederates had not planted any visible obstacles in front of the land face, such as abatises, chevaux-de-frise, or palisades. As soon as their men encountered the first torpedoes, both men realized their opponent had opted for hidden explosive devices instead of visible obstacles. Brooks was highly critical of this decision. He noted that the Confederates denied themselves the opportunity to conduct vigorous sorties at night to disrupt Union sapping operations because the minefield posed a threat to themselves. The passage through the field would have been slow and dangerous, nullifying the chief advantage of a sortie—surprise. The Confederates never attempted to come out and dispute Union progress during the siege approaches. The torpedoes "were a defense to us as well as to the enemy," Gillmore concluded.[27]

TORPEDO (ENEMY'S).

Fig. 12.

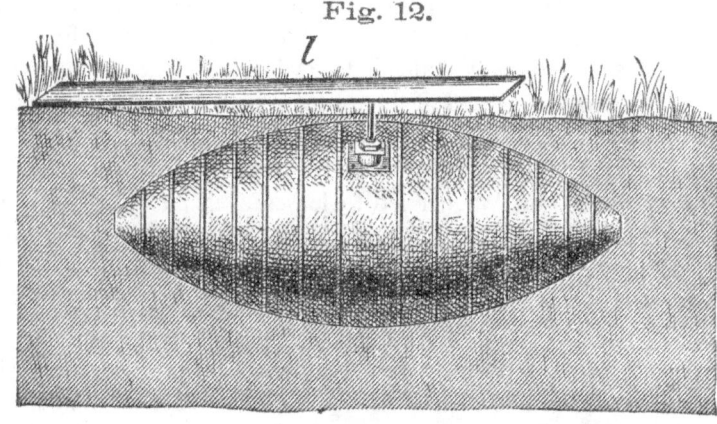

ELEVATION.

Scale 1/20 of Full Size.

Fig. 13.

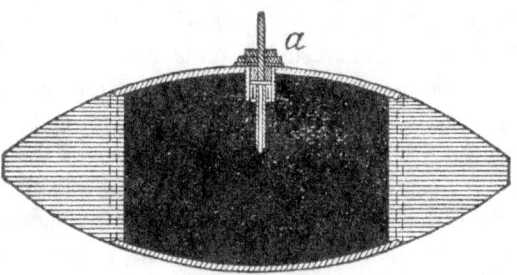

SECTION.

Figure 5.4. Fort Wagner, Keg Torpedoes. The Federals studied the keg torpedoes they found while demining the approaches to Fort Wagner and reported their findings through illustrations such as this one. (*OR*, vol. 28, pt. 1, 309)

In fact, the Confederates had considered a counterattack immediately after losing control of the sand ridge on August 26 but decided not to attempt it. The primary reason was "the want of information relative to the positions of the torpedoes in our front," wrote Col. George P. Harrison Jr., who was then in charge of Fort Wagner. Other considerations included the reported strength of the Federals on the ridge and the fact that his men would have had to approach them moving by the flank and forming line within very close range of the Union position. "These torpedoes give us considerable trouble and anxiety," reported Brooks, "but they are an excellent obstacle to prevent a sortie by the enemy, who are very much afraid of them."[28]

The minefield inhibited Confederate moves in other ways too. Engineers wanted to place spiked planks in the bottom of the ditch fronting Wagner's land face on the night of August 16 but cancelled plans when everyone realized that Gray had not kept careful records to show exactly where he had placed his torpedoes. Apparently, the engineers worked this problem out because five days later they laid down the spiked planks.[29]

As we have seen, the torpedoes failed to prevent the Federals from sapping their way forward and nearing the ditch of Fort Wagner. An infantry assault coming out of the sap was imminent, and there appeared little hope the Confederates could stop it. They evacuated Fort Wagner on the night of September 6, 1863. Federal soldiers were delighted they did not have to assault the work. Even with advancing along a sap, it is possible that the action might have spilled out onto the minefield.[30]

The torpedo danger did not end with the fall of Wagner. Up to forty-four landmines were still in place, and many Federals were eager to see the fort they had worked so hard to reduce. Col. Joseph R. Hawley of the Seventh Connecticut established guards "to check straggling" at the fort, yet quite a few men wandered around as sightseers. Half a dozen of them were injured by torpedoes during Hawley's tour of duty as officer of the day; three of them died of their injuries. Surgeon Gross reported that he treated five men for torpedo injuries after the fall of Wagner. They had tangled with landmines from September 8 through September 12, and artificer John H. Westervelt of the First New York Engineers saw an infantryman who tripped a torpedo on the night of September 13.[31]

Some Confederate officers were rather cocky about the minefield at Fort Wagner. Beauregard bragged to a Confederate senator that "I have . . . planted the ground in front of the battery with a certain kind of grain, which is quite prolific of results, even in the most barren soil." As we have seen, Beauregard's expectations fell far short of the tactical reality.[32]

Rains once again leaped at every opportunity to prove his invention had been successful. Reports in Northern newspapers indicated that the attack of July 18 had been repelled by Confederate hand grenades. Rains immediately took credit for the repulse. He knew there were no hand grenades in the fort that day and assumed the explosions were caused by the Rains shells planted by Gray on July 11. With no more "proof" than that, Rains informed Seddon that he was responsible for the saving of Fort Wagner. Seddon passed this on to Davis, who offered his gratitude at "the success of General Rains' subterra shells." John B. Jones was ready to believe anything Rains said and accepted the idea that his friend had saved the fort.[33]

Federal reaction to the use of torpedoes at Fort Wagner was muted, probably because they no longer were such a novelty by this stage of the war and because relatively few of their comrades were hurt by them. Engineers like Gillmore and Brooks examined them as a professional duty, interested in how they were made and how they were used on the tactical level, while the rank and file continued to call them "infernal machines." Brig. Gen. George H. Gordon gazed upon the stockpile of deactivated mines at the Union engineer depot after the fall of Wagner. "Killing by mines and torpedoes, not for defense, but for butchery, is scarcely civilized warfare," he concluded.[34]

FORT ESPERANZA

On the other side of the Confederacy from South Carolina, another torpedo episode developed a couple of months after the fall of Fort Wagner. Union forces made their first incursion on Texas soil in the autumn of 1863 as a way to threaten the French, who were intervening in a civil war in Mexico, and to cut trade between Mexicans and Texans that supported the Confederate war effort. For that purpose the major part of the Federal Thirteenth Corps under Maj. Gen. Napoleon J. T. Dana landed unopposed on the coast near the mouth of the Rio Grande River and then extended its control for some distance up the valley. Dana also sent Maj. Gen. Cadwallader C. Washburn north to occupy several locations along the Gulf coast, including Matagorda Island, 175 miles north of the mouth of the Rio Grande and 150 miles south of Galveston.[35]

Confederate Fort Esperanza was located on the north end of Matagorda Island, a long coastal island shielding twenty-five miles of the Texas seaboard. Pass Cavallo separated the north end of the island from the south end of Matagorda Peninsula, and just inside the near junction of these two long and narrow land features was Matagorda Bay, nestled between the barrier island and the mainland. Vessels entering Pass Cavallo could steam across the southern reaches of Matagorda Bay and into Lavaca Bay to reach two important port cities, Indianola and Lavaca, both of which had rail connections with the interior of Texas. Capt. Daniel D. Shea's Artillery Battalion (also designated the Fourth Texas Artillery Battalion) initiated the construction of

Fort Esperanza on the end of Matagorda Island early in 1862 and garrisoned it for a long while after completion. Edgar C. Singer, a member of the battalion, later created a special group of operatives whose task was to construct submarine batteries for Confederate use. Variously called the Singer Secret Service Corps, the Singer Submarine Corps, or Singer's Torpedo Company, it was organized at Lavaca early in the spring of 1863 with thirty members. The group worked up seven different types of underwater mines and other explosive devices.[36]

Singer's group began mining the waters near Fort Esperanza and those at other points along the Texas coast in May 1863. When Washburn began his move against the fort, the group was called upon by Confederate authorities for help. Capt. David Bradbury took all the mines he had available, twenty-four in number. These were underwater mines converted into sub-terra devices. Bradbury "planted them all about the trenches of the fort, with strings to set them off leading into the fort." He used the same detonation device the Confederates had employed at Port Hudson, a lanyard.[37]

Washburn landed 2,800 troops on Matagorda Island two miles from Fort Esperanza on November 26, 1863. He carefully approached the work by 11:00 a.m. the next day. His scouts reported the fort to be large and well sited, and accompanying gunboats could not yet come up to support his men due to a strong northern storm front that was passing through the area. Col. William R. Bradfute held the fort with eight hundred to one thousand Confederates. Washburn decided to await developments rather than risk heavy losses by assaulting the strong work. His men endured some artillery fire while they waited and easily captured outlying trenches located four hundred yards from the fort on November 29. Federal artillery pounded Esperanza all that day, and it appeared to Bradfute that Washburn was preparing to cut his garrison off from the mainland.[38]

The Confederates therefore evacuated Fort Esperanza at 10:00 p.m., November 29. At 1:00 a.m., the magazines inside the work began to explode. The Federals waited an hour because of the danger as a total of four magazines blew up during that time. When they entered the work at 2:00 a.m., they found three more magazines intact, but by 4:00 a.m., Washburn reported that "a hot fire is raging within the fort." The new occupants simply let the fire burn out. Washburn had lost no more than two men who were killed and two who were wounded in capturing Fort Esperanza.[39]

Sometime that morning, engineer Capt. J. T. Baker discovered Bradbury's landmines. He was walking along the parapet of the fort and looking intently at the glacis when he spotted linear disturbances of the ground, which he called "decayed grass lines." By removing the grass, he found wooden troughs "through which raw hide lines were run. In following these up, I invariably came to a torpedo." Baker cut the lines and carefully seized the "brass pin and doubl[ed] it while it was in position" before he lifted the device out of place. The rawhide lanyards reached well into the fort, and operatives could have exploded the torpedoes with safety at the moment the Federals crowded near them. "If they had made an assault on us," Confederate Bradbury asserted, the torpedoes "would have been very destructive, but as it was we lost them."[40]

Baker found and took up thirteen landmines near the fort and found three others in the water near the fort. The thirteen had been placed at "the dead angles of each salient and from the West curtain face." He found them to be cylinders measuring from thirteen to twenty-one inches in diameter and fifteen to twenty-four inches long. They contained thirty-five to forty pounds of "fine diamond-grained powder." The three mines found in the water were constructed exactly the same way as the ones found on land, except that they had "an air-vessel attached." All were made of tin, and all detonated the same way. "The withdrawal of a pin passed through the striker or hammer . . . by personal force of the contact of a floating body with lines stretched across channels." In short, it was easy for Bradbury to place the Singer device, which was entirely constructed anew, rather than using artillery shells, either in water or on land.[41]

The method of placing these mines in the ground was minutely examined by Baker. He found that they were "laid upon their sides, and were braced on their resistance side by a log of wood, or else stakes driven into the ground" so that when the lanyard was tugged, the torpedo would not move in place. Bradbury had protected the mine "from clogging with sand and earth" by covering them with boards. He also had maximized the antipersonnel effect by placing "small stones" around the cases, turning them into "makeshift projectiles." The stones mixed elements of the old fougasse with the modern landmine and served the purpose of allowing water to drain more rapidly from the hole in which the torpedoes had been placed in case of rain. The mine was not much in contact with dirt as a result and would last longer underground.[42]

Washburn had captured nine Confederates while operating against Fort Esperanza, and according to one report, they were forced to help locate and dig up the mines. Rumor had it that Washburn threatened to shoot a prisoner for every Union soldier harmed by one of the devices.[43]

Baker took up only thirteen of the twenty-four torpedoes Bradbury had placed at Fort Esperanza. The First Battalion, Fourteenth Rhode Island Heavy Artillery, assumed the responsibility of finding and taking up the rest. After the regiment had organized in December 1863, the First Battalion was transported from Providence directly to Matagorda Island by February 1, 1864. Its first task in the army was the "very delicate duty to perform in removing" Bradbury's torpedoes. They were "of a domestic manufacture—something after the style of a milk pail or a milk can with a plunger fitted in the top," wrote Lt. Charles Chace.[44]

The Confederates continued to plant many submarine mines and a few landmines at selected points along the Texas shore. Fort Griffin guarded the entrance to Sabine Pass, sixty miles up the coast from Galveston. A combined Union army and navy force attacked it on September 8, 1863, and was repulsed. A month later, Col. Valery Sulakowsky, chief engineer of the District of Texas, instructed Maj. Julius Kellersberg to strengthen its defenses with torpedoes. Sulakowsky recommended the construction of a covered way (essentially, a sunken roadway) all around Fort Griffin. Just outside the covered way, Kellersberg was to place an abatis fifteen feet wide, consisting of mesquite trees. He would further sod a strip of ground six feet wide inside the line of abatis, between it and the covered way. Before placing the sod, Kellersberg was to plant thirteen-inch shells as torpedoes, each one forty feet apart from its neighbor in this strip. They were to be "connected with wire or rope attached to stakes driven in the ground, the wire to be about 4 inches above the ground and hidden from view." Sulakowsky also ordered "a double line of torpedoes" at the southwest corner of the fort. While this is an interesting description of a mine barrier, there is no clear evidence that Kellersberg actually followed through with placing it.[45]

The Singer group proved how easy it could be to design and build a torpedo from scratch that was versatile enough to be placed underwater or underground with little adjustment of its mechanism. Bradbury carefully placed these devices at Fort Esperanza to maximize their tactical effect, and they probably would have done a lot of harm if the operative jerked the lanyard at the right moment. After examining the device, Wisconsin soldier George Wheeler thought each mine would have taken out fifty Union soldiers. But, organized as a traditional fougasse, the detonating system was not victim-activated, and thus, there were many opportunities for it to fail.[46]

But the Confederate evacuation nullified the chance to use them. Because no one was hurt by the torpedoes, the Fort Esperanza torpedoes did not register morally with the Northern public. Rains never claimed credit for them, because they did not use his sensitive primer and did not hurt the Federals in any way. Fort Esperanza therefore remained an obscure episode of landmine warfare in a region many people considered a backwater of the national conflict.

Chapter 6

Sheridan's Raid, Petersburg

"These Shells Are Now Appreciated"

The Confederates greatly expanded their use of torpedoes in 1864 and enlarged the tactical aspect of their deployment as well. They not only used landmines as an adjunct to the defense of fortified positions as they had already done at Yorktown, Fort Wagner, and Fort Esperanza but spread them along routes of communication to harass Union troop movements. Rains had experimented with this concept along the Williamsburg Road in 1862 and at Jackson the following year, but he developed a more expansive vision of the concept in 1864. He saw landmines as a good defense against rapid Union cavalry raids across the countryside and massive Union infantry raids across a large theater of operations. Each year of the Civil War brought with it another idea to the developing landmine doctrine of the conflict, and Rains tended to be at the forefront of that evolving doctrine.

Rains had become the torpedo troubleshooter of the Confederate government by the early summer of 1863 in part because he promoted himself with key Rebel officials and in part because of his genuine contributions to this new mode of warfare. His main contribution had been the sensitive primer and, to a lesser degree, the concept of utilizing artillery shells as landmines. His influence on the deployment and tactical use of torpedoes was still evolving that summer as Davis sent him from one trouble spot to another, and Rains continued to think about the tactical context of his devices. He realized that they were suited for certain situations and were not appropriate for others. After his efforts to avoid responsibility for the Yorktown landmines, he now tried to take credit for any sign that torpedoes had an effect on the Federals, no matter how slim the evidence. Under Davis's protection, Rains had in fact made himself the most important torpedo man in America.

But Rains's influence was not universally felt. He mostly had the opportunity to deal with landmines in the East and had little impact on the West and the Trans-Mississippi. In those regions, as we have seen, other torpedo men were at work independently of Rains. They were doing their job quietly, like the operatives of the Singer group, and just as effectively as Rains. Like those Singer operatives, Rains was interested in underwater mines as well as land devices. Thus, Davis sent him to coastal ports, where he consulted with engineer officers and department commanders. Rains also proposed a system of landmines designed to alert Confederates of Union cavalry raids and became heavily involved in the deployment of the most concentrated system of torpedoes defending a fortification during the long Petersburg campaign.

About two weeks after the Confederate evacuation of Jackson, Davis sent Rains to Mobile so the general could engage "in the especial duties assigned him." The Richmond authorities were careful not to mention torpedoes in their official orders, for security purposes. Sometime after August 3, 1863, the date that Johnston's headquarters issued orders for the transfer, Rains arrived at Mobile and consulted with Maj. Gen. Dabney H. Maury, commander of the Department of the Gulf. Mobile was one of the major blockade-running ports of the Confederacy, located at the northwestern corner of a huge bay, the entrance of which consisted of a narrow water passage connecting it with the Gulf of Mexico. The area was well suited for underwater mines. Even so, Gillmore's siege of Fort Wagner very soon led Davis to send Rains to Charleston less than three weeks after his arrival at Mobile.[1]

Maury was happy to see him leave. "General Rains has gone away with his gimcracks," he told Johnston on August 24. "He is not at all practical; everything I received from him was vague and visionary. He was ordered here about a week and did not commence work. The President ordered him to Charleston."[2]

Rains was not the only torpedo man in the Confederacy who was labeled a crackpot. Capt. Zere McDaniel was far less well-known than Rains but just as enthusiastic about the new explosive devices. He specialized in underwater mines and was responsible for the laying of torpedoes in the lower Yazoo River that sank the ironclad gunboat USS *Cairo* on December 12, 1862. His devices also sank the USS *Baron DeKalb* at Yazoo City on July 13, 1863. Before those successes McDaniel seemed to be just another crackpot to many who came in touch with him. "He is nearly deranged upon the subject of torpedoes," wrote Serg. Maj. William D. Elder of the First Mississippi Light Artillery. "He says that with his torpedoes he can whip a fleet of Yankee Gun boats and all that is necessary to prevent the Yankees from coming up the Yazoo River is to send them word that Capt. McDaniel is ready for them." Elder did not believe these predictions, and neither did his comrades. "Our boys have hearty laughs at the expense of the Gallant Captain and his torpedoes." Col. William T. Withers, commander of the First Mississippi Light Artillery, thought the scheme reminded him of old stories about catching birds by putting salt on their tails.[3]

These views of McDaniel and Rains were not shared by the officials in Richmond who were desperate to use any tactic to offset their disadvantages in the war. Davis hurried Rains to Charleston, instructing him to order whatever material he needed. As we have seen, Rains was unable to do anything about stopping Gillmore's sapping operations, and there already were capable men at Charleston to handle the underwater mining operations. Nevertheless, Rains remained there from August 1863 until mid-February 1864. He accumulated a pretty large staff of operatives at Charleston, building the nucleus of an administrative structure to prosecute torpedo warfare. By mid-January he had as many as twenty-two men working for him in various capacities. Some were officers, some enlisted men, and others were "hired men." Two operatives, an officer and an enlisted man, were assigned to the Army of Tennessee in Georgia, but most of Rains's men were working at and near Charleston.[4]

On February 15, 1864, Maury reported signs that the Federals were preparing a major offensive against Mobile. He requested more men and ammunition. Davis sent Rains to him once again, but this time Maury did not complain. Rains supervised the laying of many underwater mines, one of which is credited with sinking the monitor USS *Tecumseh* during the battle of Mobile Bay on August 5, 1864.[5]

While Maury did not at this second visit write critically of Rains, at least one of his subordinates found the torpedo man a strange character. "Gen'l Rains . . . was a most comical looking person," artillery Lt. Robert Tarleton wrote his friend Sallie. "You would have enjoyed him." Rains had "a short, quick, jerky way of walking, and a most peculiar expression of face, the eyes and mouth of which are incessantly in motion. Like Uncle Ned he has very little 'hair on the top of his head in the place where hair ought to grow.' He is a great torpedo man, you remember, is a perfect monomaniac on the subject—talks of nothing else. I saw him get one innocent and confiding young man in a corner and I am certain he torpedoed him for at least two hours."[6]

Despite skeptical views such as this, Rains actually accomplished something to aid in the defense of Mobile during the spring of 1864. At the same time, his patrons in Richmond helped to protect his status as the leading torpedo man of the Confederacy. Col. Alfred L. Rives, who was in charge of the Engineer Bureau, consulted with Confederate Secretary of War Seddon about a "sub terra rifle battery" proposed by Lt. John E. Sullivan of Company K, Fourth Georgia. Seddon did not think the plans submitted by Sullivan constituted a "new invention," because the principles were already well-known and already employed by Rains and other operatives.[7]

Moreover, Seddon refused to issue general directives about when and how such devices should be used. He preferred to leave it up to commanding officers at all levels to decide when it was appropriate to deploy landmines.

Whether Seddon shoved off the responsibility onto local commanders because of the moral implications involved or because those commanders were the best judges of the tactical effectiveness of these devices is not clear.[8]

Rains, in fact, spent a good deal of time and energy studying the proper tactical context for the use of landmines and, as we have seen, worried little about the moral implications. Before leaving Charleston he had proposed to Seddon a plan for placing torpedoes as a defense against Federal cavalry raids. They should be planted on all the roads leading to an important city whenever a raid was anticipated. Priming the shells could be delayed until the last minute to minimize harm to civilians in the area. "I am confident that if the enemy are once or twice blown up by these means, raids ever thereafter will be prevented."[9]

Once again, Rains assumed a lot of tactical benefit would accrue from his devices, but Seddon was doubtful. He recommended that the plan be implemented "at least to some extent" as an experiment. Davis, in contrast, was very enthusiastic about it. "I have confidence in the efficacy of sub-terra shells against cavalry under the circumstances indicated," he wrote. The army could not assign men permanently to the duty of laying and maintaining these mines, but they could be detailed temporarily to such work.[10]

SHERIDAN'S RAID

Rains proposed this plan and Seddon and Davis approved it in late September 1863, two weeks after the fall of Fort Wagner. It is not known where or when the scheme was put into operation, but during the early phase of the Overland Campaign in Virginia in mid-May 1864, a Union cavalry raid encountered landmines on roads north of Richmond. Less than a week after the start of Lt. Gen. Ulysses S. Grant's bloody campaign, he authorized Maj. Gen. Philip Sheridan to take the cavalry of the Army of the Potomac on an independent operation to engage the cavalry of the Army of Northern Virginia. Sheridan started on May 9 with 10,000 troopers and met Maj. Gen. James E. B. Stuart's 4,500 Confederate horsemen in battle at Yellow Tavern on May 11. The Federals defeated their opponents, Stuart was mortally wounded, and Sheridan continued to advance toward Richmond that evening.[11]

Brig. Gen. James H. Wilson's Third Division led the way south along Brook Turnpike and across the Chickahominy River until within five miles of the Confederate capital. Along the way, according to Wilson, at least one torpedo planted in Brook Turnpike went off when a horseman tripped it. The historian of the First Massachusetts Cavalry, however, contended that the Federals encountered many torpedoes not on Brook Turnpike but on a country road the column took to the left, or east of the pike. This byroad crossed the Virginia Central Railroad and linked with the Mechanicsville Turnpike. None of the other numerous accounts of this episode indicated exactly where the landmines were found, but all agreed there were several of them and the Federals had to deal with the devices in the middle of the night, from 11:00 p.m. of May 11 until 3:00 a.m. of May 12. Moreover, the sky was dark and rain fell off and on during this time.[12]

The landmines were planted along both sides of the roadway with wires that stretched across the middle and connected two of them in pairs. Obviously, this had been arranged only a short time before the Federals arrived in the area because it would have prevented the Confederates and any civilians from using the roadway. The devices consisted of artillery shells with friction primers. Horses disturbed the wires and set them off. Sheridan reported losing several horses and "a few men" to torpedoes that night.[13]

Capt. George B. Sanford of the First US Cavalry saw at close range the effect of these explosions. He rode with regimental commander Capt. Nelson B. Sweitzer at the front of the unit, when four horsemen rode up and crowded their way into the column just in front. Sweitzer "began to remonstrate with the party . . . ; but before he had spoken three words one of the luckless troopers rode on to a concealed torpedo and the whole set of fours went up into the air." Their horses absorbed the brunt of the explosion and saved them; none of the four men were killed in the mishap.[14]

Wilson's division carried the burden of encountering these mines during that dark and stressful night. The rest of Sheridan's column, following behind, periodically heard the torpedoes go off. Many men assumed it was Confederate artillery, only later learning that their comrades had been harassed by torpedoes.[15]

Some men in Sheridan's other divisions encountered these devices after daylight on May 12. Orderly Sgt. Wells A. Bushnell turned onto the byroad with the Sixth Ohio Cavalry after dawn, and soon "an explosion took place in the road." It did not do much harm to anyone, but his comrades investigated and discovered "a small cord connecting with several torpedoes." They followed it "into a piece of woods where a man had been stationed to pull on the cord." The Federals made the best of it, cutting the cord into short lengths and distributing it. Bushnell used a piece to secure his revolver by tying one end to his belt and the other to the butt of the weapon so he would not drop it while riding.[16]

Fear of torpedoes led to an amusing incident that relieved a moment of tension for men in the First US Cavalry. They were keenly on the lookout for landmines after the sun rose on May 12 and saw an iron bar lying on the roadway. Naturally, they assumed it was part of a booby trap. When a "blundering Irish orderly" rode by, the men shouted at him to beware. He was "bewildered" by the noise and rode on. His horse kicked the bar and it flew twenty feet away, but nothing happened. While other horses had tripped several torpedoes, this one proved to be no booby trap, and the men laughed and shouted even louder to release pent-up stress.[17]

The Federals took time on May 12 to dispose of these torpedoes. Sheridan had twenty-five prisoners and forced them to search for the devices in the early morning hours. He made them "get down on their knees, feel for the wires in the darkness follow them up and unearth the shells." One of these prisoners was Pvt. Samuel H. Nowlin of the Fifth Virginia Cavalry. The torpedoes had made the Federals angry, and Nowlin later did not blame them for forcing the prisoners to do something about the devices. "This kind of work required a delicate touch and was unpleasantly exciting," he wrote after the war. Nowlin did not indicate whether any prisoners were hurt, but the Federals enjoyed watching them at work. "Their timid groping and shrinking became a curious and rather entertaining sight," wrote Chaplain Henry R. Pyne of the First New Jersey Cavalry. According to another chaplain, Samuel L. Gracey of the Sixth Pennsylvania Cavalry, the experience convinced at least some of the prisoners that landmines were "an ignoble system of warfare."[18]

The prisoners likely played a trick on the Federals when they told Sheridan that a nearby householder was the primary person responsible for planting the torpedoes at this location. How that could be is difficult to understand, but Sheridan took it seriously. He ordered the unearthed torpedoes to be placed in the cellar of the house and rigged them to explode if disturbed. The lady of the house bitterly protested but failed to move the angry general. Sheridan also took her and everyone else at the house into custody and released them later that day. The circumstances of this story indicate that Sheridan knew the family was not responsible for the mining, but he probably assumed the house had been used by the Confederates as their base while laying the torpedoes. He wanted to surprise them when they returned and took the civilians away so they could not warn the Confederates.[19]

Sheridan's prisoners did not find and remove all the landmines placed along the roads leading to Richmond. On May 31, more than two weeks after the May 11–12 incident, a Confederate column encountered one of these devices while moving across the Virginia Central Railroad toward Cold Harbor. It had stopped to rest a bit near Ellerson's Mill. A man belonging to Capt. Philip P. Johnston's First Stuart Horse Artillery Battery saw a wire on the side of the road and pulled on it out of curiosity. The torpedo exploded a few yards away "between the lead horses" of a team in the battery. It did no damage except forcing the horses to turn so sharply that they broke the pole of the limber. It was a narrow miss and demonstrated the danger of leaving devices behind long after their usefulness had ended.[20]

On May 10, the day before Sheridan's men encountered torpedoes, Davis ordered Rains to leave Mobile and come to Richmond unless he was doing "very important" work. Apparently, he was indeed engaged in "very important work" because it took nearly a month for orders assigning him to supervise "all duties of torpedoes" at the capital to be issued. That included landmines and underwater devices for the entire Confederacy. All operatives were to report to him, all assignments for duty issued by him, and necessary material was to be accumulated and issued by his staff.[21]

But the centralized system Davis tried to create did not work smoothly. Little more than a month later, Lee complained about the lack of progress in placing underwater mines in the James River. There were places along the north bank suitable for planting them, but Rains had no boats or material to support the effort; neither did the Confederate navy. "There seemed to be some conflict of authority between the subterra and the submarine departments, which caused such delay and difficulties," Lee observed, that work stopped. He thought there was ample opportunity to play havoc with Union transports plying the river if the mines could be placed at key positions. "It matters little which department wins the credit, provided the work is well done," he concluded. Lee's understanding was that the James River fell into Rains's jurisdiction, but Rains did not seem to be willing to do anything.[22]

The problem lay in part with typical Confederate ineptitude in administrative matters and probably with Rains himself. The general does not seem to have been a good administrator, and he was effective only when his personal touch could be felt in a particular locality. Shortages of material generally plagued the Confederate war effort, but Rains's lack of energy and his inability to remain in touch with key commanders in the field had nothing to do with material shortages.

By October 1864 Rains exhibited signs of greater activity. At Lee's request, he dispatched some torpedoes and men able to deploy them to Col. John S. Mosby, commander of a partisan unit in the Shenandoah Valley. He also sent torpedoes and operatives to Goldsborough and Wilmington, North Carolina, when requests from area commanders arrived. The navy asked for landmines to guard land approaches to their installations, and Rains responded. "These shells are now appreciated," he told Seddon on October 29, "and I now have more calls for their use than I can possibly fulfill." Rains also had dispatched all his available operatives. He asked Seddon for authority to create a school for officers so he could train adequate numbers of men "to carry our efforts into effect."[23]

Rains felt he was in his element, encouraged by the increasing demands for his invention and now in charge of the Confederacy's torpedo efforts. But once again, his influence was mostly felt in the East, reflecting the constricting nature of central authority in the shrinking Confederacy. Nevertheless, he continued to grasp at any indication that torpedoes were effective. A friend of his from Mobile relayed information that USS *Tecumseh* was not sunk by gunfire from Fort Morgan during the Battle of Mobile Bay but by an underwater mine. Rains was careful to let Davis know of this information.[24]

By late November Rains reported more fully to Seddon his limited success in mining the rivers of Virginia and North Carolina. He used mostly the Singer underwater mine and dispatched Dr. John R. Fretwell, an expert in its deployment, to work in the field. He indicated that after one success with the mines on the James River, the Federals instituted effective countermeasures. They deployed guards to watch segments of the river bank and thus restricted Rains's ability to place more devices. By this time the Confederates had mines in the James, Pamunkey, Chickahominy, and Appomattox Rivers of Virginia and were planning to put some in the Tar River of North Carolina. But Rains was unable to supply more operatives for Charleston, the center of underwater mine warfare on the Atlantic coast.[25]

Meanwhile, Rains had developed a way to apply his sensitive primer to the existing hand grenade to make it more effective. Called, for some reason, the dart grenade, it was "about the size of a goose egg," according to Brig. Gen. Edward Porter Alexander, who called the primer "a sensitive paper percussion fuze" inserted into the forward end. Another innovation was to tie a two-foot-long cord to the rear end. "A man could swing one of these & throw it 60 yards, & they would burst wherever they struck." Alexander ordered "a large lot" of them made for close work against Union troops at Elliott's Salient during the Petersburg campaign. Here the opposing fortified lines were no more than 125 yards apart, and Alexander suspected the Federals would try to sap across no man's land.[26]

PETERSBURG

The most visible evidence of Rains's work appeared along another sector of the Petersburg lines in the fall of 1864. On October 10 Davis urged him to confer with commanders about the use of torpedoes as a defensive measure. Rains went the next day and "indoctrinated them into the use of the subterra shells. They now gladly avail themselves of this means of defense." In fact, six hundred devices had been planted in front of the works by October 27, and sixty more were ready to be placed on October 28.[27]

The impetus for this spurt of activity was Grant's Fifth Offensive at Petersburg from September 29 to October 2, which resulted in the capture of Confederate Fort Harrison, north of the James River. The Federals ruptured but did not break the Rebel line, converting Harrison into Union Fort Burnham as the Confederates adjusted their position to accommodate the Federal success. In the wake of this near catastrophe Rebel commanders were ready for any aid to their defensive measures.[28]

The landmines placed at Petersburg had been prepared months before. Zere McDaniel had been busy fitting 3,953 "wooden fuze plugs to 53,352 pounds of condemned shells" in January 1864. Those projectiles included 3,233 24-pounder shells, 381 Parrott 150-pounder shells, and 141 8-inch mortar shells.[29]

Operatives planted these projectiles from Fort Gilmer to Elliott's Salient (not to be confused with the Elliott's Salient south of the Appomattox River that had been the target of the famous Crater Battle of July 30). They placed the devices eighteen inches, two feet, or five feet apart, depending on which source one relies on. The landmines were planted between a line of stakes that was located close to the ditch of the earthwork and the abatis that lay farther out. Some sources indicated they were placed in a single line, while others claimed there were two lines of torpedoes between the ditch and the abatis.[30]

In his "Torpedo Book" of 1874, Rains described not only the generic method of planting landmines but specifically referred to the Petersburg minefield he helped to create. After preparing the shells, he recommended they be transported to the site in caissons, ambulances, spring wagons, or "in bags on mules if necessary." After digging a hole, the operative should attach a tin shield over the shell, using chalk to mark on the shells how far down the shield ought to rest on it. Rains recommended that the bottom part of the shield be cut into scallops to fit the configuration of the shell as closely as possible, making a fairly watertight covering over the top of it. Then it was a matter of removing the screw plug from the fuse hole, inserting his sensitive primer, fixing the shield over it, and planting the device. Rains knew this delicate operation was dangerous. He advised everyone to carefully select men "of staid & sober habits," and to drill them in the process before arriving at the site "to prevent the possibility of accident."[31]

Rains explained his device in more detail in the pages of the unpublished "Torpedo Book" than in any other writing. The sensitive primer had "a cylindrical cover on top made of tin" with "paper wet in mucilage wrapped around it, and an additional short cylinder of thick leather pushed upon the shank by it." He had added the leather piece "to allow the tin cup cover, when trodden upon, to be forced down upon the cone of the primer to explode it." The tin shield needed to be flexible enough so that pushing on its top would bend the metal and apply enough pressure to touch off the mine. Before that time, it was the best waterproof cover for the device. Rains coated the tin shield in coal tar and rolled it in the sand "to take on the appearance of the earth as much as possible."[32]

If a tin shield was not available, placing a board or shingle on the ground resting over the primer would work. Rains thought an oil cloth, India rubber cloth, or tarred paper would suffice to keep the primer dry in wet weather. If there was nothing available that could be used as waterproofing, Rains suggested that coating the sensitive primer "in a solution of gutta percha in chloroform repeatedly and a thick coat of tallow put on its under side" could keep it dry for some time.[33]

Another innovation by Rains was the placing of warning flags in the minefield. A small red flannel flag, ten inches square, tied to a three-foot-tall stake was placed three feet behind each mine. They were to be taken out each evening and replaced every morning. Two larger flags marked the location of walkways through the field, and Rains devised a lantern with a red glass shield to use at night, when Confederate pickets had to pass through the mine belt. In his "Torpedo Book," Rains suggested that the ten-inch red flags were placed to warn the Federals as well as the Confederates. He felt so certain that the mere presence of his devices would terrorize the enemy that there was no need to conceal their location.[34]

The Petersburg planting was the first time that Rains placed torpedoes in a line, and he admitted it was done to minimize the exposure of friendly troops to their danger. The elaborate flag arrangement was implemented for the same reason. All this was to counter "objections to these subterra shells," as he told Seddon. Rains saw the mines as strengthening the defense of this sector so that it could be held by "inferior troops," freeing good men "for more active service in the field." In fact, this sector of the Confederate line was held in part by three battalions of City Troops, which were placed near Fort Gilmer. Brig. Gen. Seth M. Barton's Division of the Richmond Forces held the rest of the sector that was mined. It consisted of Brig. Gen. Patrick T. Moore's Local Defense Brigade and Barton's City Brigade, commanded by Col. Meriwether Lewis Clark.[35]

"These shells now seem to be popular with our officers," Rains exulted in a dispatch to Seddon on November 18. He was supplying them as fast as possible to the tune of about one hundred every day. "From reports of deserters they are rapidly demoralizing the enemy." By November 18 Rains reported having planted a total of 1,298 torpedoes. Apparently, their placement continued after that date, for Rains claimed in his "Torpedo Book" of 1874 that a total of 2,363 landmines had been planted north of the James River that fall.[36]

The distance between Fort Gilmer and Elliott's Salient was 2,266 yards. If 1,298 mines had been placed eighteen inches apart, they would have created a line 649 yards long. If all of them had been placed five feet apart, the torpedo line would have been 2,163 yards long. This planting was by far the biggest torpedo field yet created in the Civil War, and it was possibly larger than the unknown number of Russian mines deployed at Sebastopol. We do not know the pattern used by the Russians in 1854, but in the Civil War, the Petersburg minefield was the first to be arranged in a linear formation and the first to be marked by flags. Rains had solved the problem of waterproofing his devices

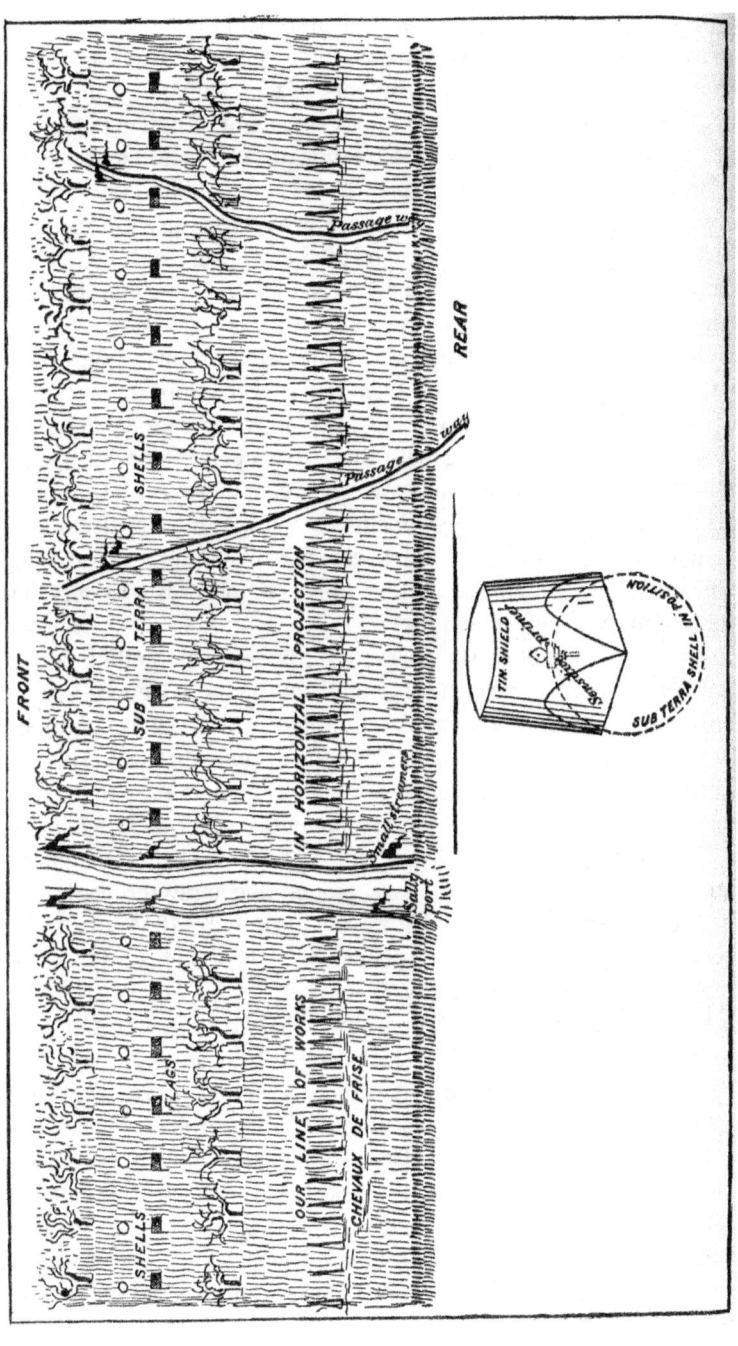

Figure 6.1. Petersburg. This illustration, drawn by Rains or his subordinates, demonstrates the first true mine belt in the Civil War. It was a well-organized field of torpedoes that stretched just in front of 2,266 yards of the Confederate Petersburg-Richmond line north of the James River. By November 18, 1864, Rains had planted between 1,298 and 2,363 torpedoes here, creating the largest mine field yet seen in the Civil War, the first to be arranged in linear fashion, and the first to be marked by flags. The illustration graphically supported Rains's doctrine of landmine use. (*OR*, vol. 42, pt. 3, 1221)

and had taken seriously the question of danger to his own people. His concept of torpedo warfare had advanced a great deal since Yorktown.

Confederate soldiers were fully aware of the landmines at Petersburg and described them in their personal accounts of the campaign. Sgt. Carter Nelson Berkeley Minor of Company I, First Confederate Engineers recalled that some ladies from Richmond visited the lines in late October to see the torpedoes. They brought a large dog along that became rambunctious and ran into the minefield. This incident caused a stir as everyone tried to coax the dog back without tripping a device. One of the ladies even told the soldiers to shoot the dog if necessary, but it managed to come back without touching off a torpedo.[37]

The Federals found out about the mines almost as soon as they were planted. An agent behind Confederate lines reported the appearance of torpedoes in front of Fort Gilmer, covered with planks and marked by small flags. He found this out by talking with the Rebel operatives who planted the devices. Confederate deserters also were an accurate source of information. Elizabeth Van Lew, a Union sympathizer and spy in Richmond, sent word with a deserter that the Confederates were also planting more torpedoes on the roads leading to Richmond similar to those Sheridan's cavalry had encountered five months before.[38]

Rains's precautions did not entirely prevent his people from getting hurt by the torpedoes. In November one of his operatives, William S. Deupree, "accidentally fell upon one and was immediately killed in full sight of the foe." The Federals came to the site to observe "the effects and what was doing."[39]

A Confederate infantryman became a victim of landmines on another part of the Petersburg lines. South of the Appomattox River, near where the Norfolk and Petersburg Railroad crossed the Rebel line, Federal infantrymen had advanced a sap to advantageous ground and threw grenades at a Confederate picket post only twenty yards away on July 22, 1864. Brig. Gen. Archibald Gracie recommended that torpedoes be planted on his brigade sector as a result of this aggressive activity.[40]

There is no documentation concerning how, when, or where this suggestion was acted upon, except that Surg. Jack Bryant Stinson wrote an article published after the war about a torpedo casualty among Gracie's men at Petersburg. The landmines had been placed on the right of way and between the ties in a line across the railroad. Confederate pickets were warned of their location and told not to go near the spot. But Sgt. Jerry Burkhalter of Company D, Forty-first Alabama ignored the warning one day. He believed he could step on all the ties to avoid trouble but missed one and trod on a torpedo. Burkhalter's leg was badly mangled, but he refused to let the surgeon amputate and died about ten days later.[41]

Torpedoes were planted at a third location along the Richmond-Petersburg lines as well. Late in the campaign, Longstreet wanted to place "some sensitive shells . . . among the abatis in front of our lines, especially opposite" Fort Burnham. The general wanted the devices to go off if the Federals tried to tear the abatis apart in preparation for an attack. Longstreet had come a long way from his initial reluctance to condone the use of torpedoes in May 1862. Also, there are indications that the line of torpedoes extended from Elliott's Salient all the way to the James River by March 1865.[42]

The Confederates created a torpedo belt along the Richmond-Petersburg lines that was the most carefully planned barrier to Union assault to be seen not only in the Civil War but also in world history. By advertising their existence, Rains made sure the Federals were fully aware of the danger. Col. Edward Hastings Ripley, an Eighteenth Corps brigade commander, described to his father "the formidable barriers of red earth, abattis, double and triple rows of which divide us, with torpedoes and 'trip irons' well sprinkled in." The soldiers knew more than did the newspapers or the civilians at home why it seemed to be taking so long to break the apparent stalemate at Petersburg. "If any insane gent sitting slipshod at home reading the papers, wonders why Grant didn't advance more on Richmond, the 18th Corps would like the job of taking him out on our picket lines, and tossing him atop one of their infernal machines to show him exactly why."[43]

As the campaign neared an end in late March 1865, Maj. Gen. Edward O. C. Ord issued instructions for his subordinates in the Army of the James about how to deal with the torpedo belt in case the Confederates evacuated Richmond. "Look out for torpedoes and mines," he wrote. "Don't let your columns take the roads, keep them in the woods and bypaths. Send cattle and old horses up the roads first."[44]

The Federals made no effort to plant landmines in front of their works along the Richmond-Petersburg lines, because Lee rarely conducted attacks against them. The Union army and the Northern populace had maintained their disgust with the use of antipersonnel mines throughout the conflict, which is another reason for this neglect of the new defensive weapon. But Union engineer officers examined the torpedo belt after the fall of Richmond and Petersburg and filed reports documenting the technical aspects of it. They estimated that only seven pounds of pressure were needed to touch one off and also noted that Rebel operatives had, at times, placed boards across several of them to maximize the effect if a Federal soldier stepped on the plank.[45]

The Confederate torpedo belt north of the James River represented a new step toward modern deployment of landmines in its ordered nature and its mass of devices planted within a constricted space. It is true that the

Federals never contemplated an attempt to penetrate it, but that was mainly because Grant gave up further attacks north of the James River after the Fifth Offensive. Rains's carefully placed barrier was never tested.

Chapter 7

March to the Sea, Pooler Station, Fort McAllister

"This Was Not War, but Murder, and It Made Me Very Angry"

Toward the latter part of 1864, the Confederates intensified their use of landmines, as their hope for survival became more desperate. Especially in theaters where their troop strength was low and the Federals were numerous, they saw torpedoes as an equalizer both to defend cities and forts and to delay the movement of Union columns. As we have seen in Virginia, all previous doubts about the use of these devices evaporated among the Confederates in the Western campaigns of late 1864 as well.

MARCH TO THE SEA

Sherman's campaign for Atlanta, lasting from early May until the fall of the city on September 2, 1864, was the major event of the Western war that year. After four long months of continuous campaigning, his large army group compelled Gen. John Bell Hood's Army of Tennessee to give up the city. Hood retired to Lovejoy's Station, twenty miles south of Atlanta, where Sherman confronted him for a few days before breaking off contact and retiring to Atlanta for rest. Now free to initiate strategic moves of his own, Hood shifted his troops to Palmetto Station, twenty-four miles southwest of Atlanta, to plan an offensive against Sherman's railroad supply line with Chattanooga.

Jefferson Davis visited the army at Palmetto Station and consulted with Hood about the next move. Their immediate concern was that Sherman would advance toward Augusta, where Col. George W. Rains operated a major gunpowder factory. Davis stopped at Augusta after leaving the Army of Tennessee to inspect the city's defenses and contacted Gabriel J. Rains to prepare landmines for planting along the roads leading to the place.[1]

It is not known whether Rains was able to do this, but Hood made his move while Sherman's command was still resting at Atlanta. Moving out of Palmetto Station in late September, the Army of Tennessee crossed the Chattahoochee River and struck at the Western and Atlantic Railroad, breaking it in several places. Sherman brought most of his army group north to give chase in what came to be known as Hood's North Georgia Campaign. He was unable to catch the Confederate army, and no large battle ensued. After inflicting some damage on the line, Hood moved northwest into the northern part of Alabama to prepare for an invasion of Middle Tennessee. Sherman called off further pursuit and returned to Atlanta to prepare for a divergent move into the Deep South after detaching enough troops to hold Nashville. The two major concentrations of Union and Confederate manpower in the West were preparing to march in opposite directions.

Leading sixty thousand hardened veterans, Sherman left Atlanta on November 15 and headed not for Augusta but generally toward the southeast. Nothing stood in his way except a division of cavalry led by Maj. Gen. Joseph Wheeler. Confederate authorities became desperate. P. G. T. Beauregard, recently appointed to superintend military operations in the West, issued a call to the people of Georgia three days later to "obstruct and destroy all roads in Sherman's front, flank, and rear, and his army will soon starve in your midst!" Beauregard instructed Wheeler to harass Sherman's advance and delay his progress while destroying all supplies in his path. Davis reiterated these instructions to his aide-de-camp, Col. William M. Browne, who was at Augusta at the time. All commanders in Georgia were to hinder Federal movement "by destroying bridges, felling trees, planting sub-terra shells and otherwise."[2]

Confederate authorities saw landmines as an impediment to Sherman's strategic move, and Beauregard mobilized the weapon. He telegraphed Seddon from Corinth, Mississippi, to send "a large supply of Rains' subterra shells, with [a] competent person to employ them," to Macon, a city that seemed to be on Sherman's itinerary. When Seddon showed this telegram to Davis, the Confederate president instructed his staff to find out if George W. Rains had a supply of his brother's landmines in storage at Augusta. The answer was yes, so Davis telegraphed Maj. Gen. Howell Cobb at Macon that he could call on Rains at Augusta for the devices, which he was certain would "be effective to check an advance." Beauregard also found out about the stored torpedoes at Augusta and telegraphed Cobb about them.[3]

A few days later on November 21, Maj. Gen. Samuel Jones at Charleston requested the services of James H. Tomb, the chief engineer on board the *CSS Chicora*, an ironclad ram in Charleston Harbor. Tomb had earlier told Jones that he was familiar with the use of torpedoes and the road system of Georgia and was certain he could do something to impede Sherman's movement. Jones wanted to give Tomb an opportunity to try and requested Flag Officer John R. Tucker's permission to detach him from the *Chicora* for temporary duty on land. Tucker readily assented, and Jones contacted Lt. Gen. William J. Hardee at Savannah that Tomb was on his way "with some torpedoes, which he thinks he can use to some advantage." Jones told Tomb to report to Hardee and "explain the nature of the service on which he is sent."[4]

By early December it became increasingly clear that Savannah was Sherman's objective. Beauregard was at Augusta by then and telegraphed to subordinates in the area to be active in blocking roads in Sherman's path by felling trees and planting landmines. "I recommend torpedoes where roads and fords are obstructed," he told Hardee.[5]

POOLER STATION

Thus far, the frenzy of telegrams had not resulted in any blocking of roads or planting of torpedoes since Sherman had left Atlanta. But as the Federal columns neared Savannah, Hardee sent Tomb out to place torpedoes along one road the Unionists were bound to use. Pooler Station, also known as Station No. 1, was located on the Georgia Central Railroad nine miles west of Savannah. A wagon road ran roughly parallel and a short distance north of the track, traversing a cypress swamp near the station. Early on December 9, Tomb planted torpedoes in the road just where it entered the swamp, which appeared impassable due to the depth of the morass. It was a choke point for Federal movement.[6]

Brig. Gen. Joseph H. Lewis's Brigade of Kentucky mounted infantry, detached from Wheeler's cavalry, held the road. Johnny Green of the Ninth Kentucky observed as Tomb and his assistants planted torpedoes on the road, which was planked at this point because of the low-lying ground. They lifted up some planks and put the devices under them. Lewis established his line of battle a quarter of a mile east of this spot, where the road exited the swamp, and waited for the Federals to approach.[7]

The van of one column of Sherman's command neared Pooler Station along the wagon road later that day, December 9, 1864. Leading the column was a regiment of Southern loyalists, the First Alabama Cavalry. Organized in the summer of 1862, it had mostly seen service in the backwaters of the war until assigned to Maj. Gen. Frank P. Blair's Seventeenth Corps late in October 1864. The regiment was pushing Lewis's men back two miles on the morning of December 9 as Tomb planted his torpedoes at the edge of the swamp.[8]

Blair had assigned Lt. Col. Andrew Hickenlooper, his assistant inspector general, to coordinate the cavalry advance of the Seventeenth Corps. Hickenlooper had come to realize that the adjutant of the First Alabama Cavalry was the key to its operations. Lt. Francis W. Tupper, born in New York but living in Illinois at the outbreak of the war, had enlisted as a private in the Fifteenth Illinois Cavalry in 1862. He was able to transfer to the First Alabama Cavalry with a commission as lieutenant in November 1863. Tupper "was the master spirit—as brave, gallant and dashing a soldier as ever straddled a horse," wrote Hickenlooper. He had worked closely with Tupper since the start of the campaign and had developed "a strong feeling of friendship" with the lieutenant.[9]

December 9 was cloudy and chilly with weather that was "raw and ugly all day," according to Maj. Henry Hitchcock, assistant adjutant general on Sherman's staff. Hickenlooper and Tupper rode together at the head of the First Alabama Cavalry until they reached a slight ascent in the generally even ground. From here they could see the road stretching straight as an arrow toward the cypress swamp. A group of Confederate horsemen, members of Lewis's Brigade, were near where Tomb had planted his torpedoes and displayed a "generally defiant attitude," as Hickenlooper put it. He and Tupper consulted and decided to push them out of the way.[10]

The First Alabama Cavalry drew sabers and charged down the road first at a trot and then a gallop. The Confederates scattered before contact was made, but the momentum of the Union horsemen carried them onto the torpedoes. Tomb had planted nine devices in a line across the road with four directly in the roadbed and the other five on the right of way. The two located in the center of the roadbed were touched off by horses' hooves. One of the first to arrive on the scene was Capt. John C. Van Duzer of the US Military Telegraph, who was attached to Sherman's headquarters. Van Duzer saw one horse with a foot entirely cut off, but its rider was not hurt. Other troopers were not so lucky. Reportedly, six men were wounded and six horses killed by the explosion of the two torpedoes.[11]

By the time Van Duzer arrived on the scene, Hickenlooper and Tupper were already trying to find more landmines. Neither had been injured in the initial riding over the torpedo line. "I could see the black heads of several of the torpedoes sticking up above the road surface," Hickenlooper recalled, because five of them had been planted in the right of way rather than under the planks of the road surface. He found a stick and began to scrape away the dirt from around one plunger. "Suddenly, came a crash and I was blown over into the road. In a moment I regained my composure enough to wipe the dust from my eyes and look around."[12]

Unfortunately, Tupper had stepped on one of the two mines yet planted in the roadbed. As the lieutenant had stood examining the scene, he saw a fragment of one of the two exploded mines and moved his right foot in order to stoop over and pick it up. That foot touched off one of the two remaining mines that were hidden by the planks in the roadbed. Van Duzer was standing eight feet from Tupper when this happened and claimed that Hickenlooper was even closer. Neither man was injured, but a shell fragment hit the heel of Tupper's right foot and plunged through the lower part of his leg until exiting near the knee. The lower leg was terribly mangled.[13]

As soon as he wiped the dust from his eyes, Hickenlooper "saw poor Tupper lying upon his side with his [right] leg blown off just below the knee. I hastily dragged him to the side of the road and applied a tournequet [*sic*], which I always carried, to stop the flow of blood." Hickenlooper sent his orderly to find a surgeon. Tupper gradually became conscious as they waited for medical help and asked Hickenlooper where he was injured. "I told him his leg was gone but he insisted he was hurt somewhere else. I could find no other wound. He was absolutely insensible to the real injury."[14]

Sherman and several of his staff members soon rode up and were saddened by the scene. Tupper lay under a blanket, his face very pale, still waiting for the surgeon. An intact torpedo with the plunger clearly visible above the sand was not far away. Sherman spoke briefly with Tupper, and someone lifted the blanket so the general could see "the terrible wound." Sherman made further inquiries about the particulars of the incident. "There had been no resistance at that point," he related in his memoirs, "nothing to give warning of danger." The torpedoes had been planted so as to be hidden under the planks, giving the Federals little chance to become aware of them before tripping the devices. "This was not war, but murder," Sherman concluded, "and it made me very angry."[15]

Soon after Sherman and his staff arrived at the scene of Tupper's injury, a handful of Confederate prisoners neared the area. Blair ordered them to begin the process of finding and bringing up the landmines. "They were greatly alarmed," wrote Hitchcock in his diary. "Two of them begged Gen. S. very hard to be let off." Sherman refused. "He told them their people had put these things there to assassinate our men instead of fighting them fair, and they must remove them; and if *they* got blown up he didn't care. He did exactly right." The prisoners set about to look for more mines, and Sherman "could hardly help laughing at their stepping so gingerly along the road." They found the six remaining torpedoes and dug them up without incident, placing them fifty feet from the road near a tree.[16]

Word of the torpedoes soon spread through the ranks, often with embellishments to make the story even more dramatic. One report had it that Sherman consented when a Confederate prisoner suggested he be allowed to return to Rebel lines and tell everyone that the Federals used prisoners to dig up mines.

Figure 7.1. Lt. Francis Tupper. Adjutant of the First Alabama Cavalry (US), Tupper was severely injured by a Confederate landmine at Pooler Station on December 9, 1864. In letters to his family, he became the only victim of a landmine in the Civil War to describe the incident and his recovery from the amputation of a leg. (www.1stalabamacavalryusv.com/Roster/Troopers.aspx?trooperid=2341)

According to another rumor, Sherman forced the prisoners to march over the roadbed in order to set off any remaining torpedoes.[17]

No one was better at remembering wild rumors in his memoirs than Lt. John Henry Otto of the Twenty-first Wisconsin. According to him, the incident happened at the van of the Fourteenth Corps rather than Blair's

Seventeenth Corps, and Maj. Gen. Jefferson C. Davis was responsible for forcing prisoners to march over the road. Two of them were killed and several injured. The Federals sent a message to Hardee, recalled Otto, informing him of how the prisoners were used as mine clearers. He responded by insisting that he knew nothing of these torpedoes and would make sure whoever was responsible would be punished.[18]

None of these rumors were based on facts, but they reflect the emotional reaction to the Pooler Station incident among Union soldiers. Everyone felt that Sherman was fully justified to use captured Confederates to clear the torpedoes even if it resulted in the death or injury of the captives. Otto's wild story about Hardee apologizing for the torpedoes and promising to bring Tomb to justice was the ultimate in wishful thinking. It apparently consoled him to believe that at least one high-ranking Confederate officer had enough sense of pride to condemn the use of landmines in a situation such as this.

At the scene of Tupper's injury, Hitchcock took the time to examine the excavated mines. He found that several were twelve-pounder shells "with a sort of nipple projecting from the fuse-hole." Others were "large copper cylinders, rounded at each end." One of these underwater mines measured thirteen inches long and seven inches in diameter. Two brass nuts were screwed onto one end, supporting "some sort of friction tube." Hitchcock estimated each of the underwater mines held four to five pounds of gunpowder. He did not know for sure what the three mines that had exploded consisted of, but it appears that Tupper had stepped on an artillery shell because a large fragment had done most of the damage to his leg.[19]

The torpedoes failed to do more than delay Union operations a short while. After firing artillery for an hour, Blair ordered Maj. Gen. Joseph A. Mower's First Division, Seventeenth Corps to deal with the situation. Mower positioned Brig. Gen. John W. Sprague's Second Brigade across the road to hold the attention of Lewis's Kentuckians while he sent his other two brigades, led by Brig. Gen. John W. Fuller and Col. John Tillson, through the cypress swamp to the south of the road. The flanking troops had to wade through stagnant water knee-deep but eventually made it through the swamp and drove Lewis's men from their position before the end of day on December 9.[20]

The Federals were not impressed by the planting of these torpedoes, from a tactical viewpoint. They recognized that the position was one well suited to a stubborn defense, with a narrow road straight through a swamp that constituted a choke point in the Union advance. The landmines could not stop that advance, but a more determined resistance by Lewis's Brigade could have caused more trouble. "General says rebs ought to have made more of it," Hitchcock wrote of Sherman. Hitchcock himself agreed with his chief. "I was surprised that rebs had made no more effectual defense at this swamp, for it seemed to me an ugly place to pass through."[21]

Fenwick Y. Hedley of the Thirty-second Illinois thought these torpedoes had not been placed primarily to kill Federal soldiers but to serve as a warning of their approach. Given that the main Confederate line was some distance east, on the other side of the swamp, the Rebels had limited visibility. It is true that Confederate artillery with Lewis's Brigade opened fire at the sound of those two landmine explosions. One shell burst near the color company of the Thirty-second Illinois, which was part of Brig. Gen. William W. Belknap's First Brigade of Brig. Gen. Giles A. Smith's Fourth Division, Seventeenth Corps. It wounded the captain and five men of that company, but no one was killed. The regiment retired to the north side of the road and into the tree cover as a result of that round, which took out about as many Federals as had the two torpedoes.[22]

Francis Tupper became the first commissioned officer to be hurt by a landmine in the Civil War, and Hickenlooper undoubtedly saved his life by applying a tourniquet so soon after the injury. Soon after Sherman and staff had arrived on the scene and just before the Confederate prisoners were forced to search for torpedoes, an ambulance and a surgeon arrived for the stricken lieutenant. Tupper was transported three hundred yards to a house just inside the woods bordering the swamp and to the north side of the road. Sherman and staff also stopped there and Hitchcock inadvertently caught a glimpse of Tupper just before his leg was amputated. It was "horribly torn and mutilated, raw and bloody end of bone and torn muscles, etc., where torn off. Piece of shell had also run up leg *inside* along the bone, and came out near knee, shattering bone." The surgeon cut off the limb above the knee. Some doubts about Tupper's chances of recovery were expressed, but Hitchcock was hopeful. Two or three other wounded Federals, probably also victims of the Pooler Station torpedoes, occupied the house as well.[23]

African Americans had been drawn to Sherman's moving columns by the thousands during the March to the Sea, and many of them had warned the Federals to be aware of torpedoes planted along the roads. They obviously had heard of those frenzied Rebel efforts to obstruct Union progress. But those warnings did not prevent the Federals from reacting with bitter anger toward the Pooler Station incident. Members of Sherman's staff took their cue from their commander. Hitchcock called the Confederates "cowardly villains." It conceivably was justified to plant torpedoes in front of a fort because the presence of the earthwork itself was a kind of warning of their presence. But to plant them under the planks in a public road with no warning sign made of these landmines little less than "murderous instruments of assassination—contrary to every rule of civilized warfare." Like Sherman, Hitchcock was disgusted that the Confederates did not "stand and fight like men" at Pooler Station. "These rebs are certainly insane," Hitchcock concluded, "their torpedoes could at most kill a few of our advance guard—could not possibly delay or interfere with such an army—and they *know* must inevitably exasperate our men!" Capt. George Ward Nichols, one of Sherman's aides-de-camp, echoed Hitchcock's views in his memoirs. But these views were not restricted to Sherman's staff. They were widely shared by commentators throughout the ranks.[24]

FORT MCALLISTER

Pooler Station was but the first encounter with torpedoes for Sherman's army group during the March to the Sea. As the Federals neared Savannah, they found a city ringed with earthworks to protect the land approaches as well as three permanent or semipermanent forts to guard the seaward approaches to the port city. While Sherman's men had lived high on the rich produce of central Georgia during two-thirds of their march from Atlanta, they encountered food shortages during the last one-third of the way as they traversed the sandy soil of the coastal plain. Confederate Savannah, garrisoned by ten thousand troops under Hardee, stood between Sherman and the Union navy. He was eager to regain contact with the outside world and open a line of supply.

The best way to do that was to capture Fort McAllister, a large earthen fort located nine miles up the Ogeechee River and about twelve miles southwest of Savannah. Mounting twenty-two guns, most of which were mounted on the river face, the work was designed to stop Federal warships from ascending the Ogeechee and had already repelled one attack in 1863. Fort McAllister was detached from the land defenses of Savannah and thus vulnerable to a quick assault. The few guns mounted on the land face were placed on barbettes, and the crews would be exposed to rifle fire.[25]

Maj. George W. Anderson's small garrison at Fort McAllister had ample warning of Sherman's approach and worked hard to prepare its defense. "A short time before" the Federals arrived in the area, recalled Anderson, "a member of the torpedo department" came with "a considerable number of sub-terra shells" and planted them along the land face and on roads approaching the fort. This was not done by Anderson's engineer, Capt. Thomas S. White, and must have been the work of James H. Tomb.[26]

According to Capt. Orlando M. Poe, Sherman's chief engineer, the land face had a ditch with a palisade planted in the bottom and a row of good abatis a few feet out from the ditch. Tomb planted his mines in a row just outside the abatis. They consisted of eight-inch shells placed three feet from each other with percussion fuses in every one, according to Poe. Another officer, Lt. Col. William E. Strong of Maj. Gen. Oliver O. Howard's staff, Army of the Tennessee, asserted that Tomb had planted 150 thirteen-inch shells along the land face. Brig. Gen. William B. Hazen, whose Second Division, Fifteenth

Corps was assigned to capture the fort, noted that the torpedoes were located one hundred yards out from the abatis. In all of the conflicting details about the location and type of mines used at Fort McAllister, Poe and Hazen should be given priority because of their ready access to the details of the story.[27]

Tomb also placed torpedoes on a narrow causeway leading toward the fort from the west. It was located between the Ogeechee River and a salt marsh to the south. Here is where Hazen initially encountered landmines. Lt. Henry Bremfoerder and a squad of men from the Forty-seventh Ohio preceded Hazen's division and captured a Confederate outpost on the causeway about a mile from the fort at 11:00 a. m. of December 13. When marching the captives to Hazen, Bremfoerder noticed that the Confederates filed to the extreme edge of the road. He stopped and asked them why and was astounded to learn that torpedoes were planted in the roadway. Bremfoerder compelled the Rebels to search for and dig up at least some of them. Hazen was informed and he placed guards where any landmines were left in the ground, notified his brigade leaders of the danger, and waited until the Confederates completed their task of unearthing the devices.[28]

Bremfoerder's capture of these Confederates and their willingness to divulge the location of the mines saved Hazen a good deal of trouble and some casualties. Avoiding the danger, he moved his division to the nearby Middleton House about a mile from the fort and organized for the assault. Hazen placed eight regiments in reserve near the spot where Bremfoerder had captured the picket and arrayed nine regiments in a thin, single-rank line to attack the work. He hoped that small arms and artillery fire as well as the landmines would be less effective on such a thin formation. The line formed six hundred yards from the fort as sharpshooters deployed to harass Confederate artillery crews.[29]

Hazen's attack started at 4:45 p.m. The Federals lost few, if any, men before reaching the torpedo belt, and then they began to suffer more casualties. They tripped landmines, "blowing many men to atoms," Hazen reported, "but the line moved on without checking." Theophilus M. Magaw of the Forty-seventh Ohio reported that "some men" were torn "all to pieces."[30]

An unidentified member of the reserve force recalled the experience with some degree of color nearly thirty years later. "All of a sudden there was a peculiar sound of explosives, and we saw men leap into the air and fall and a little wavering in the line, but it was only momentary, as on they went again." As his regiment followed up the attacking line, the men could see the killed and wounded at the torpedo belt. "The scene was awful, and our men were horribly mutilated, for they had been torpedoed and literally blown to pieces." According to this man, the Confederates had left small flags made of flannel to mark the location of the torpedo belt. The Federals did not notice them, their gaze fixed on the fort beyond.[31]

Figure 7.2. Fort McAllister. The attack by Hazen's division on December 13, 1864, produced a number of torpedo casualties. According to available statistics, landmines accounted for 14.1 percent of the wounds treated in hospitals after this attack. The mines continued to exact casualties even after the short battle, until they could be taken up. (*Harper's Weekly*, January 14, 1865, 20)

On the Federal right, the Thirtieth Ohio, Sixth Missouri, and 116th Illinois from Col. Theodore Jones's First Brigade moved along the river bank, negotiating mud and entanglements until reaching a point only twenty yards from the fort. Here the men encountered the torpedo belt. Several troops were injured in the Thirtieth Ohio, including Sgt. Lyman Hardman. "I exploded a torpedo that had been placed in the ground by stepping on it," he later recalled. "On recovering from the effects of the shock I found that the shoe of my left foot was blown entirely off and the foot very badly burned. My eyes swelled shut in a short time. The sufferings of that night were terrible."[32]

Once past the torpedo belt, the Federals made short work of the job. They crossed the ditch despite the palisade, clambered up the parapet, and fired into the defenders at point blank range. In a moment the attackers cleared the land face of Confederates and then entered the fort. Most of Anderson's men tried to find refuge in bombproofs but were rousted out by the Federals who fired through the entrances. By 5:00 p.m., only fifteen minutes after starting the attack, Fort McAllister was under Union control.[33]

There was some discrepancy in reports of losses during this brief battle. Asst. Surg. David L. Huntington, Howard's acting medical director, stated that Hazen had lost twelve killed and 80 wounded, but Hazen himself placed the toll at twenty-four killed and 110 wounded. Many commentators asserted that the landmines accounted for the majority of the deaths and injuries. The Forty-eighth Illinois lost both its color bearers to torpedoes, and Huntington reported that "the wounds thus inflicted were generally of a grave nature."[34]

But an accurate count of the wounds suffered by Hazen's men was preserved and donated to the US Sanitary Commission after the war by Surg. John Moore, Sherman's chief medical director. It indicated that a total of ninety-two men wounded at Fort McAllister were treated in the army's hospitals. Of that number, fifty-two were injured by rifle balls (56.5 percent); eighteen by shells (19.5 percent); thirteen by torpedoes (14.1 percent); eight by pistol balls and buckshots (.08 percent); and one by bayonet (.001 percent). The report did not take into account the men who were killed on the field, but among the wounded who received hospital treatment, it is clear that the majority were injured by small arms fire and by artillery to a greater proportion than by torpedoes.[35]

But it is interesting that everyone assumed landmines had done more harm than anything else in the assault. That is a testament to the emotional impact of these mines rather than to their tactical effectiveness. Fourteen percent of the treated injuries that were caused by torpedoes is a relatively high proportion, considering the very limited numbers of Federals who had been injured in previous landmine incidents from Yorktown to Fort McAllister, but that limited level of effectiveness failed to even delay Hazen's attack, much less stop it.

Figure 7.3. Clearing Torpedoes at Fort McAllister. The Federals used Confederate prisoners to help clear the mines at Fort McAllister. Two torpedoes have already been dug up; note the boards used to expand the area of contact. (*Frank Leslie's Illustrated,* January 21, 1865, 276)

Sherman and Howard visited the captured fort soon after its fall. As they neared the place, a sentinel "cautioned [them] to be very careful, as the ground outside the fort was full of torpedoes." Dead men still lay about inside the work as the two surveyed the scene. A soldier searching for a missing comrade stepped on a landmine outside the fort and was torn to pieces. An ambulance also hit one of the devices. "Mules, ambulance, and men were blown into the air," wrote Howard. Maj. Thomas W. Osborn, Howard's artillery chief, confirmed that a number of Federals tripped landmines and became casualties after the fort was captured.[36]

In the aftermath, the Unionists did what they often had done since the start of these torpedo incidents—force captured Confederates to clear the mines. Sherman instructed Anderson's engineer, Thomas White, to supervise sixteen prisoners in this work, which began on December 14. Anderson thoroughly disapproved of this but was powerless to prevent it. After the war he called it "an unwarrantable and improper treatment of prisoners of war." Rear-Admiral John A. Dahlgren, commander of the South Atlantic Blockading Squadron, also assigned his staff members to supervise the process. As the mines were being uncovered, the admiral noted that many of them consisted of seven-inch shells.[37]

The torpedo incidents at Pooler Station and Fort McAllister led to an overwhelming feeling of anger and disgust in Sherman, which was reflected in less obvious ways among his subordinates. Anderson recalled years later that his initial meeting with Sherman a short time after the fall of the fort was testy. In "a furious voice," the general asked him, "Do you condone the use of torpedoes in civilized warfare?" Anderson sidestepped the question by telling him, "I was sent to Fort McAllister to obey orders, not to question them." That, of course, did not soften Sherman's rage. "It's inhuman. It's barbarous. And this is your Southern Chivalry." The unidentified soldier in Hazen's division, quoted earlier, noted that his comrades did not speak loudly their view. But "there were many vows of retaliation for what we considered fiendish warfare."[38]

Neither the passage of time since Yorktown nor the repeated Confederate use of landmines had dulled Northern anger at their employment. But the repeated use had offered Federal observers the opportunity to evaluate their tactical effectiveness. The verdict was one of only partial vindication of Rains's high hopes for the weapon. Henry Hitchcock talked to some Confederates captured at Fort McAllister, and they were "thunderstruck" at the quick success of Hazen's attack. "They say that they had confidently expected our line to be repulsed by the torpedoes alone."[39]

The reason this did not happen was that the torpedo belt was too thin to have much effect on the attacking line, which Hazen had spread out into something like a skirmish line, instead of a shoulder-to-shoulder formation. Depth in the mine field was the key, a lesson not yet learned by Confederate operatives.

But the emotional effect of torpedoes was more to be reckoned with than the actual destruction they caused. In that sense Rains was right. This point also was recognized by many Federals. "The usefulness of torpedoes . . . is very questionable," concluded engineer officer William Ludlow soon after the war. He admitted that planting them in the bottom of the ditch or under the abatis could serve as a useful tactical ploy but their emotional impact was more important. "They operate in all cases as much by their moral effect as by actual destruction of life. Men are very cautious of treading where the presence of a torpedo is suspected."[40]

The Federals came across more torpedoes before the confrontation at Savannah ended. Sherman positioned his army group fronting the land defenses of the city and then tried to extend his line to cut off the only avenue of retreat open to Hardee, north of the Savannah River into South Carolina. Just before completing that process, Hardee evacuated the city without a fight

Figure 7.4. Torpedo Found at Savannah. Maj. Gen. Absalom Baird donated this landmine to the US Military Academy. It was found among a row of landmines stretched across the Louisville Road, which linked the town of Queensboro at the head of navigation on the Ogeechee River to Savannah. The row of torpedoes lay in front of the land defenses of Savannah. The unusual design suggests it was initially intended for some other purpose than a sub-terra torpedo. (USMA and Herbert M. Schiller, *Confederate Torpedoes*)

on the night of December 20–21, 1864. On the left wing of the Union line, between the Georgia Central Railroad and the Savannah River, Capt. Hartwell Osborn of the Fifty-fifth Ohio was serving as the officer of the day for Col. Samuel Ross's Third Brigade, Brig. Gen. William T. Ward's Third Division in the Twentieth Corps. When he heard that Federal troops on another part of the line had entered the Confederate works at 6:00 a.m., he pushed Ross's skirmish line forward to do the same. Six Confederates gave themselves up, "pointing out at the same time three torpedoes which had been imbedded in the bed of the railroad." Ross's brigade line advanced with its right at the railroad, avoiding the spot indicated by the deserters.[41]

To the right of the Twentieth Corps, south of the Georgia Central Railroad, two Black men told the Federals about landmines. They confirmed that the Confederates had evacuated the defenses of Savannah during the previous night and "had left a few torpedoes planted on the railroad to our left," in the words of men who served in Battery I, Second Illinois Light Artillery.[42]

Landmines had played a prominent role in the mind of Federal troops operating near Savannah, although they failed to influence the tactical flow of events. But, as engineer Ludlow concluded soon after, there was ample evidence that underwater mines were largely effective in dealing with Union naval vessels. "It is in the defence of harbors against a hostile fleet that torpedoes find their true place and pre-eminent utility," he wrote. The Federals found this to be the case at Savannah. Their opponents had planted numerous underwater mines in the Savannah and Ogeechee Rivers, which continued to block Union supply shipments to Sherman's men after the fall of the city. In the Savannah, a double row of piles offered even more of a hindrance to navigation. Ludlow saw some underwater mines left behind in Savannah by the Confederates. They were yet unfinished but consisted of a powder barrel with wooden caps on each end extending to a point. The Rebels had covered everything with tin and applied a coat of boiled tar to waterproof the weapon. Exploding devices consisted of nipples attached to percussion fuses. Anchored at key locations, they were effective antiship weapons.[43]

During the confrontation at Savannah, Francis W. Tupper experienced a relatively fast recovery from his terrible injury suffered at Pooler Station. He was not only the first commissioned officer to become a victim of a torpedo in the war, but he thoroughly documented his injury and recovery in letters to his parents in Illinois.

Following the amputation of Tupper's right leg in the small house only three hundred yards from the site of the injury, Hickenlooper took personal responsibility for his friend's care. He arranged for an ambulance to carry him elsewhere because the Federals did not establish a field hospital in the area. Tended by an assistant surgeon, Tupper remained in the ambulance until arrangements could be made to transport him to Beaufort, South Carolina, where he ended up in the officers' general hospital by about December 19. On his departure for Beaufort, Hickenlooper recalled that Tupper was "on the high road to complete recovery." The two friends parted forever because the circumstances of their lives never again intersected.[44]

"I met with an accident that has made me a cripple for life," Tupper informed his parents on Christmas Eve. "My clothes were literally torn in rags a piece [of the torpedo] cut through my right sleeve to my shirt without touching the arm." Although the stump of his right leg was healing well, Tupper felt the emotional effect of his injury. "The loss of my limb is a severe one to me and I expect that you Father & Mother will feel it heavily, but you must take everything as it comes. The most wonderful part of the whole is that I wasn't instantly killed."[45]

Despite depression over the loss of his leg, Tupper kept up his spirits as best he could. The hospital was located in a large house on Main Street, near the bay, and conditions were not crowded. Tupper was in an upstairs room with four other patients. The place was well organized, clean, and staffed by professional doctors and nurses. A surgeon visited him every morning and the stump was washed and dressed every day. Tupper endured pain but otherwise enjoyed good health. He lay in bed for several days after arriving, sitting up now and then to write letters using the hard surface of a book cover as his writing table.[46]

Tupper spent a quiet Christmas and New Year's, only sending out for mince pie to celebrate the coming of 1865. Soon after, he did not want for visitors. When elements of the Army of the Tennessee moved through Beaufort in the early stages of the Carolinas campaign, he received many calls from well-wishers. They included Oliver O. Howard, who had lost his right arm due to a wound received at the Battle of Fair Oaks in June 1862.[47]

Another amputee lodged in a separate room of the hospital also paid a visit of sorts with Tupper. He began to yell loudly one day, "Hallo you one legged adjutant." Tupper responded, and the man told him that he had lost his left foot and wanted to arrange with Tupper for the two to buy their boots together. Tupper had not yet seen him as of January 15, because neither could leave their beds for the time being. But he learned he was "a large strong constitutioned fellow and full of life. He sings hollers preaches and keeps that ward in an uproar all the time."[48]

Tupper's stump continued to heal rapidly, and the hospital staff assured him it would accommodate an artificial limb very well. "After some practice I can walk readily without cane or crutches." He felt certain members of his regiment, the First Alabama Cavalry, would pitch in with donations to purchase "one of the best of manufactured limbs" for him. Before the end of January, the persistent pain had lessened considerably.[49]

The first time he tried to walk, on February 9, two months after the injury, Tupper had some difficulty. He mounted a pair of crutches and began to move about in the room. "It made my leg bleed a little from the small veins. The Dr did not like this move much. I put up the crutches and told him that I would admire them for a while longer before using them." The surgeon thought it would take another two weeks for the stump to heal over, and then it would be possible to walk on the crutches.[50]

The landmine Tomb planted under the planks in the road at Pooler Station did not prevent Francis Tupper from rebuilding his life. His stump healed, and he recovered as much as one could expect from the loss of a leg. Tupper moved west and became county clerk at Denver, Colorado. He married and led a largely normal life while dealing with the inevitable limitations imposed by having to use only one leg.[51]

Chapter 8

Fort Fisher, Sister's Ferry, Carolinas

"This Low and Mean Spirit of Warfare"

The Confederates continued to hope that landmines could help their desperate efforts to stave off defeat in a losing war during the early months of 1865. Again, they used them to support the defense of semipermanent fortifications, as well as to impede the march of Sherman's veterans through the Carolinas. Once again, a handful of Federals were injured but not dissuaded from their tactical missions. Fort Fisher on the Atlantic coast fell despite a system of electrically detonated torpedoes. Sherman's men encountered landmines at the start of their march north from Savannah, but the episode only reinforced their bitter view that the Rebels were barbaric in nature. Rains's invention and his tactical doctrine for its use once again produced results far below his overheated expectations.

FORT FISHER

Wilmington, North Carolina, located several miles up the Cape Fear River, had emerged as the most important blockade-running port in the Confederacy by the latter stages of the war. Material shipped from there went directly to Robert E. Lee's Army of Northern Virginia, penned in its fortifications at Petersburg and Richmond. The main protection of Wilmington was Fort Fisher, located near the end of Federal Point, a strip of coastal land that formed the north bank of the Cape Fear as it emptied into the Atlantic Ocean. The fort was begun in April 1861, but Col. William Lamb led a massive effort to enlarge and strengthen the work in 1862. More than two years later, by the fall of 1864, it was one of the largest semipermanent earthworks in the Confederacy, with a long sea face and a shorter land face that was 1,300 yards long. Consisting of huge sand mounds mounting heavy artillery, which were connected by curtains, Fort Fisher was a formidable target for any assaulting force.[1]

The first rumor of a Federal attack arrived on October 24, 1864, and Lamb responded by planting a row of torpedoes five hundred to six hundred feet in front of the land face. It took his operatives two months to complete the electrically detonated torpedo belt.[2]

But the first Federal expedition against Fort Fisher was a dismal failure, and that was not because of the landmines. Troops from the Army of the James, led by Maj. Gen. Benjamin Butler, were detached from the Petersburg-Richmond lines and transported by sea. Grant loaned his chief engineer, Lt. Cyrus B. Comstock, to the expedition, which was supported by sixty-four warships under Rear Admiral David D. Porter. On December 24 this mighty fleet arrived near Fort Fisher and opened a heavy bombardment. Butler landed his infantry on Federal Point, north of the work, the next day, but there was to be no attack. Butler had his heart set on a scheme to explode a boat loaded with gunpowder as near the fort as possible in hopes of flattening the sand mounds. The experiment proved the fallacy of such a hope. His confidence shaken, Butler evacuated his position on December 27 and retired.[3]

The torpedoes were not tested, but Lamb had an intricate plan for their use. He had hoped to lure the first line of Federals over the belt and allow them to close on the ditch of the fort's land face. He planned to then explode the torpedoes to keep the second line of infantry isolated from the first.[4] He banked heavily on the assumption that the emotional effect of the torpedoes would prevent the Unionists from even trying to cross the belt. As we have seen at Fort McAllister, that was a faint hope.

Butler's handling of the first Fort Fisher expedition was so embarrassing that he was finally relieved as commander of the Army of the James and replaced by Maj. Gen. Edward O. C. Ord. Grant wanted a second expedition mounted as soon as possible. Ord detailed Maj. Gen. Alfred H. Terry to command nearly ten thousand men, close to twice as many as Butler had taken. Porter once again supported the effort but, this time, with fifty-eight warships. Terry landed his men at the same place on January 13. The next day Porter opened an intense bombardment that continued into January 15.[5]

The bombardment had its effect. The palisade along the land face was battered and the artillery mounted on the sand mounds was largely silenced. Most importantly for our story, naval shells cut most of the electrical lines that linked the torpedoes with Fort Fisher. When the Federals went in, they would not have to contend with landmines. At this point the Federals did not know there were any torpedoes. In fact, they had dug a forward trench line quite near the devices without detecting them.[6]

The assault began at 3:45 p.m. on January 15. Lamb ordered his electrician to explode the torpedoes as soon as the Federals crossed the belt and lodged at the foot of the parapet, but the man could not do so. The Federals fought their way into the fort and engaged in close range combat with the garrison that lasted until dark. Eventually, they captured the work section by section.[7]

After the battle, the Federals discovered the landmines. Cyrus Comstock took charge of them as they were dug up by a pioneer company commanded by Lt. A. L. Knowlton of the Fourth New Hampshire. Knowlton's men excavated a total of twenty-four mines planted about two hundred yards out from the fort and fronting the land face. They were about eighty feet apart from each other, constituting "an elaborate system" of defense. Upon digging them up, Knowlton divided them into three types. First, there were two twenty-inch shells, and next, what he termed eleven "boiler-iron cylinders" that were thirteen inches in diameter and eighteen inches in length. Third, Knowlton counted eleven "buoy-shaped vessels" made of sheet iron and about the same size as the cylinders. The second and third types apparently were designed as underwater mines. All the torpedoes contained from fifty to one hundred pounds of gunpowder, depending on their size.[8]

The Confederates had rigged three sets of insulated copper wires to connect this array with firing stations in the fort. One set ran out to the mines covering the left sector of the land face, another did the same for the right sector, and the third covered the center. Each set consisted of a double line for insurance and then branched out into several single wires to reach a number of mines per set. "A single wire running to a group of torpedoes was branched to each, in the expectation, apparently, of having battery power sufficient to fire the whole group." In addition, Comstock observed that "some of these groups were connected with each other, thus giving (with sufficient battery power) a choice of positions in the work to fire the group from."[9]

Naval shells had cut both wires in the left set and right set, but the center set remained intact. This was fortunate because the infantry assault took place against the left of the land face, crossing one of the two deactivated sections of the torpedo belt. Porter had organized a force of marines and sailors to participate in the assault, and this contingent attacked the right end of the land face, exactly over the other inactive sector. If Terry had sent men in to strike the center of the land face, undoubtedly the operator would have exploded those mines when they were near or past the torpedo belt.[10]

None of the twenty-four torpedoes were detonated, mostly because of the cutting of the connecting wires. But Comstock speculated that even those in the center might not have worked if the operator had tried to set them off. He examined the fuse of one of those mines and found that the powder had caked. Comstock also had some doubt about whether the batteries found inside the fort were strong enough to have generated the power to set off the several mines to which they were attached. While he was impressed by the layout of the torpedo system, he harbored more than one reservation about its technical effectiveness.[11]

The batteries were electromagnetic, but Comstock did not indicate who had made them. "A few turns of the crank . . . with the black lead connection readily firing gunpowder in fine grains," would have exploded the mines, he told Richard Delafield, the chief engineer in Washington, DC. A photograph of what has been identified as one cell from a galvanic rather than an electromagnetic battery, found near Fort Fisher, has turned up. If so, the Confederates may well have had a chemical battery at the installation as well as electromagnetic batteries but preferred the latter when it came time to lay the mine system.[12]

Comstock closely examined the fuses, or means of deflagration, the Confederates used in the torpedoes. He found that, in many cases, they were grounded by extending a three-foot-long wire from the fuse into the ground, but in other cases, the wire was attached to the metal sheeting of the underwater mines that were used in the system. The fuse contained mealed powder, which Comstock thought could be the kind used in slow-match fuses. He provided a detailed description for Delafield, which included gutta percha, copper wires, wood, and twine. Waterproofing was accomplished with "cotton yarn, greased," but as indicated earlier, Comstock found caked powder too, which indicated the cotton yarn was inadequate. Kochan and Wideman, the most recent historians of torpedo technology, provide a photograph of an original sub-terra device used at Fort Fisher and identify the Abel magnet fuse as being used in it. It had been developed by the leading British authority on explosives, Frederick Augustus Abel, who was an ordnance chemist at the Chemical Establishment of the Royal Arsenal at Woolwich. Kochan and Wideman also identify the Wheatstone Magnetic Exploder as the battery used at Fort Fisher but do not provide a source for that information.[13]

How effective on the tactical level this sophisticated system of torpedoes might have been is a matter of conjecture. William Lamb was convinced it would have prevented the Federals from capturing Fort Fisher, especially if the palisade had remained intact. In this regard he spoke as would have Gabriel Rains if the latter had made any comment on the issue. Rains ignored the landmines at Fort Fisher because they were not of his design and he did not believe in electrical detonation. But, as we have already seen, a single line of landmines completely failed to stop Hazen's attack at Fort McAllister, and there is no reason to believe that Lamb's system would have stopped Terry's attack at Fort Fisher. Comstock assumed that the torpedoes would have caused "several losses and demoralization" but not the failure of the assault.[14]

The story of torpedoes at Fort Fisher did not end with the taking up of the system. They were implicated in a terrible disaster that struck the victorious

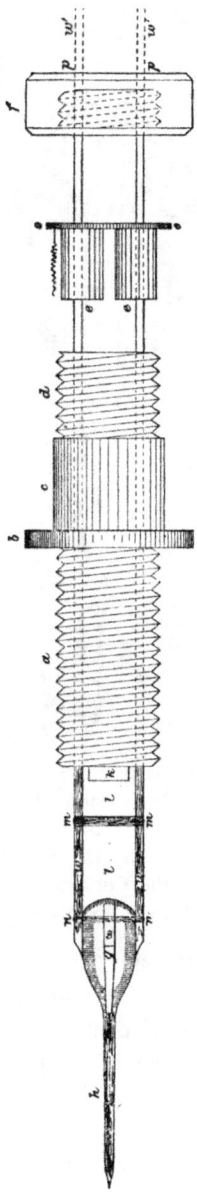

Figure 8.1. Fort Fisher Fuse. A fuse developed by Frederick Augustus Abel, an ordnance chemist at the Chemical Establishment of the Royal Arsenal at Woolwich, England, was used by the Confederates in their torpedo belt at Fort Fisher. Designed to be activated by energy from a Wheatstone electromagnetic battery, most of the lines linking the battery to the torpedoes were cut by Union artillery fire. As a result, none of the landmines were exploded by the operator when the Federals attacked and captured Fort Fisher on January 15, 1865. (*OR*, vol. 46, pt. 2, 217)

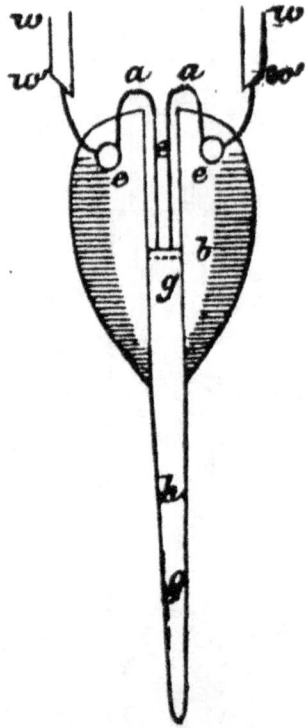

Figure 8.2. Fort Fisher Quill. This is the quill, or point of the fuse, developed by Abel. This and the entire fuse were minutely examined by Lt. Col. Cyrus B. Comstock, who filed a report on the torpedo system at Fort Fisher. Comstock also brought some samples of the Abel fuse to show Grant, on whose staff he served as chief engineer. This is the only instance we know of wherein the Confederates obtained components for their landmines from outside sources. (*OR*, vol. 46, pt. 2, 215)

Federals. Just after dawn of January 16, the morning after the battle, the main magazine of the fort suddenly exploded. It contained thirteen thousand pounds of gunpowder, and all of it went up, killing and wounding two hundred Union and Confederate soldiers. Four days later Terry appointed a court of inquiry to discover the cause of this catastrophe. Lt. Col. Nathan J. Johnson of the 115th New York testified that he reached the scene right after the explosion and noticed "two wires, one solid copper, the other, I should judge, a submarine wire, composed of seven small ones inclosed in rubber, leading from the magazine toward the Cape Fear River in a northwesterly direction, I cut both for fear of further explosions." These wires were laying on the surface of the ground for some distance but then entered underground as they neared the river.[15]

Maj. Ezra L. Walrath of the 115th New York added fuel to the fire of suspicion when he told the court that he discovered in the ruins of the magazine "the conical end of a torpedo which had burnt powder on the end and had certainly been at some time exploded." Curious, Walrath looked around and found apparently the same wires that his lieutenant colonel had discovered. He also cut the wires and reported that earth thrown up by the explosion had covered the section of wires near the magazine. The wires also were exposed for lengths of ten to fifteen feet as they ran toward the river.[16]

Word of the explosion and the evidence concerning a deliberate attempt by the Confederates to blow up the fort after it fell came to captured and wounded Rebel officers in the nearby hospital. They tried to disassociate themselves from any blame for the incident. Maj. Gen. William H. C. Whiting, who was in overall command at Wilmington, and Maj. James H. Hill, his assistant adjutant general, were in the hospital and assured Surg. Andrew J. H. Buzzell of the Third New Hampshire that the only torpedoes were those planted outside the land face of the fort. There were "no wires connected with the magazine," in Buzzell's words. Whiting had already told Terry this information after the fort fell.[17]

The court of inquiry came to the conclusion that the Confederates were not responsible for the explosion of the magazine. Its officers did not "attach any importance to the report that a magnetic wire connected this work with some work on the opposite side of the Cape Fear River." They concluded that the explosion was the result of carelessness on the part of the Federals. Although officers had posted sentinels at service magazines located in the sand mounds along both land and sea faces, they had failed to post a guard at the main magazine. As a result, anyone could and did wander into and out of it for curiosity's sake. One report had it that a group of marines had gone into the magazine with lighted torches.[18]

The court did not take seriously the evidence of wires provided by Johnson and Walrath. The probable explanation for those wires is that the Confederates had planned to deploy underwater mines in the Cape Fear River, using the main magazine as a firing station. They undoubtedly used the magazine as a storage facility for their land and underwater mines, thus explaining the fragment Walrath had discovered. Based only on the testimony of Hill and Whiting, the court concluded that there was no Confederate plot to blow up the magazine after capture.

Despite the tragic explosion, the fall of Fort Fisher closed Wilmington as the last blockade-running port in what was left of the Confederacy. Yet Rains tried to sell cotton through the blockade in order to purchase material for his torpedo campaign against the Federals. On January 28, only a week and a half after Terry captured Fort Fisher, Rains requested permission to run fifty bales of cotton through the blockade for sale at Nassau in the Bahamas. Rains listed six Wheatstone electromagnetic batteries along with one thousand primers and five miles of gutta percha cable. He included several different kinds of tin, 480 yards of "coarse red flannel" cloth, and a "double seaming machine for torpedo caps." The items amounted to more than $7,000 worth of material, but Rains insisted they were "an absolute necessity." He envisioned some sort of secret expedition behind Union lines once his operatives had possession of these items. The sale never took place. The War Department pointed out that all cotton was to be sold to the account of the Treasury Department, not to any other agency within the government. Moreover, with Wilmington out of the picture, it was very difficult to plot any course of action involving cotton exporting or purchasing war material in foreign lands.[19]

SISTER'S FERRY

Shortly after the Fort Fisher landmines failed to arrest the fall of that work, Sherman resumed his march through the heart of the Confederacy. He sent most of Howard's Army of the Tennessee (Fifteenth and Seventeenth Corps) by ship to Beaufort, South Carolina, to begin its march through the Carolinas. But Maj. Gen. Henry W. Slocum's Army of Georgia (Fourteenth and Twentieth Corps), started its march by a different route. Two divisions of the Twentieth Corps crossed the Savannah River at Savannah and marched up the South Carolina side of the stream toward the northwest. The Fourteenth Corps, one division of the Twentieth Corps, a division of the Seventeenth Corps that could not cross at Savannah, and Sherman's only cavalry division moved up the Georgia side of the river. This column included twenty thousand men and one thousand wagons. Slocum aimed to cross it at Sister's Ferry, thirty-five miles upstream from Savannah. The intention was to join the two divisions that had marched up the South Carolina side of the river at Robertsville, seven miles northwest of Sister's Ferry, and begin the march from there.[20]

Slocum's men began to cross on January 29, 1865, by laying a pontoon bridge over the swollen river. The Fifty-eighth Indiana, a regiment dedicated to taking care of the pontoons, placed 250 feet of floating bridge over the main channel and constructed 750 feet of wooden bridging to connect it with both sides. The bottomland was flooded with one to ten feet of water, and once over the main channel, the Federals had to rely on a causeway that ran parallel to the river for two and a half miles before turning inland onto higher ground. This causeway proved to be the most difficult part of the journey. The Confederates had cut trees across the roadbed, which forced details from Fourteenth and Twentieth Corps units to clear them away.[21]

Soon after starting this task on January 30, the details encountered torpedoes. They were planted in the roadbed under the felled trees and came as a shock to the men. A sergeant in the Seventy-ninth Pennsylvania stepped on one, and the explosion tore his left leg off below the knee. Surgeons cut the stump off above the knee joint. Another man was injured by the explosion. A captain nearly became a victim of a mine. He stepped on one, but it failed to detonate.[22]

After those incidents, work was stopped until officers could investigate. They soon found about fifty more landmines along the causeway. Ironically, the Confederates had carefully marked the location of all these devices because they did not know for sure if the Federals would use this route. They wanted to quickly dig up the mines and move them elsewhere if needed, so they had placed a small peg "on one side of the road and marked with the number of the torpedo" to indicate the placement of each one. They were six- and ten-pound shells, "arranged either with a plunger and cap, like a percussion shell, or with a friction arrangement, on the principle of a friction-tube," reported engineer officer William Ludlow. James Tomb probably was the chief agent behind the planting at Sister's Ferry. Word of the torpedoes quickly spread to Sherman's headquarters, which traveled with Howard's wing, and to Maj. Gen. Henry W. Halleck in Washington, DC.[23]

"Torpedoes! Be careful!" read a sign planted at the South Carolina end of the bridge. It had an effect on the soldiers crossing the pontoons. The work of clearing the causeway of landmines proceeded as quickly as was safe for the men detailed to the job. Troops from the 102nd New York not only helped engineers dig them up but also repaired the wretched roadway by laying corduroy along the causeway after the road was cleared of both torpedoes and felled trees.[24]

All the torpedo clearing at Sister's Ferry was conducted by the Federals. One postwar source indicated that Confederate prisoners were used, but it does not seem reliable. The desire to use prisoners, however, was very strong. Chaplain John J. Hight of the Fifty-eighth Indiana wanted to "make a chain gang of them—officers are preferable. Let them remove all they can find, then, let them draw heavy wagons before the advance of our army. This will effectually cure the rebels. They will soon become tired of blowing up their own people."[25]

The Confederates learned of the landmine explosions at Sister's Ferry on January 30 and knew that "some damage" had been done. But Slocum considered Tomb's torpedoes to be a nuisance on par with the flooded bottomland, the felled timber, and the bad road on the causeway. All of these obstacles delayed his advance by several days, with torpedoes playing no more than a modest role in that delay.[26]

Figure 8.3. Clearing Torpedoes at Sister's Ferry. The Federals in this illustration seem pretty satisfied with their success in finding and digging up Confederate landmines along the road leading from Sister's Ferry into South Carolina early in 1865. (*Frank Leslie's Illustrated*, March 11, 1865, 397)

While the mines failed to have much tactical effect, they once again raised bitter anger among the Federals. It was "an unfortunate introduction" to South Carolina for Slocum's men, commented Lt. Henry I. Smith of the Seventh Iowa. The Federals already were in a hostile frame of mind because South Carolina had been the first state to leave the Union. They "could not help feeling bitter against such a mode of cowardly defense." Smith acknowledged that planting torpedoes to defend an obvious position was legitimate, but to hide them on a road was "something akin to poisoning a stream of water." The state suffered more than the usual destruction during the course of Sherman's march and deserved it, as far as Smith was concerned. Brig. Gen. Jefferson C. Davis, commander of the Fourteenth Corps, agreed with Smith and concluded after the campaign that "South Carolina has since paid the penalty of a resort to this low and mean spirit of warfare."[27]

CAROLINAS CAMPAIGN

Sister's Ferry was a warning to the Federals to expect more torpedoes as they marched through the Carolinas. Maj. Gen. John G. Foster, commander of the

Department of the South, relayed reports of landmines in South Carolina to Howard. A Black man warned Foster that torpedoes had been planted near a two-gun battery located between Port Royal Ferry and Pocataligo. The former place was ten miles northwest of Beaufort and the latter an additional ten miles northwest of the ferry. Even though the Black man told Foster that the torpedoes had been known to exist months before, in August 1864, Foster warned Howard because the latter's Army of the Tennessee would be marching through that general area. Howard's men, however, did not encounter any problems.[28]

But according to John Hill Ferguson of the Tenth Illinois, members of the Seventeenth Corps came across several landmines farther north during their march toward Columbia, South Carolina. "A number of accidents," as he put it, took place on February 13. One man was killed and three were wounded in the Twenty-fifth Indiana when they investigated a potato hole and set off a landmine. Both the Tenth Illinois and the Twenty-fifth Indiana served in Col. John Tillson's Third Brigade, Maj. Gen. Joseph A. Mower's Second Division. Ferguson also noted that one man was killed and five wounded in Brig. Gen. Giles A. Smith's Fourth Division "while attempting to open a large box which they found in a swamp." While members of Brig. Gen. Manning F. Force's Third Division tore up railroad tracks near Orangeburg, they touched off a torpedo that killed and wounded two or three men.[29]

Ferguson's reports may be valid, but we have no confirmation of these torpedo incidents from other sources. Many years after the war, relic hunters found two landmines "in an old roadbed in the low country of South Carolina," according to historian Charles H. Jones. Both devices had been specially made as torpedoes, with small holes on the side to pour in gunpowder. The fuses were plunger type, but because one had a wire attached, it is possible an electrical ignition system had been rigged for it.[30]

The Confederates were making desperate efforts to plant torpedoes in Sherman's path. Rains dispatched Capt. Garnett McMillan "for duty with subterra shells." McMillan reported to Beauregard on February 12, who sent him to Lt. Gen. Wade Hampton at Columbia. Hampton ordered him to plant torpedoes on a causeway five miles south of the city, but McMillan had many difficulties fulfilling the directive. He was ready early on February 15, but Hampton's quartermaster could not spare any wagons. At noon someone found a civilian ambulance for him, but by the time he reached the spot, the Federals had secured the causeway. McMillan planted no landmines. "From Columbia to Camden has been such a field for the work that I return heart sick at nothing being done," he reported to Rains. "Much may yet be done between Camden and Raleigh, Wilmington, or what point soever the enemy may be destined for." McMillan wanted a "light, strong spring-wagon, with water-proof cover," plus "a pick or grubbing-hoe" to prosecute the work. He assured Rains that Hampton appreciated the value of these landmines and was eager for his services.[31]

Hampton may have appreciated the mines when McMillan reported to him on February 15, but he later changed his opinion. The general was not overly disappointed by McMillan's failure to mine the causeway. "The shells there would have impeded their march on that road, but would not have seriously delayed them," he wrote. "I do not think any serious damage could have been inflicted on them by these subterra shells." McMillan retained his ambulance and joined the rest of Hampton's wagon train. When Beauregard ordered the train away on February 18, McMillan went with it, taking his shells with him. Upon meeting Hampton again at Chesterfield, McMillan told the general he was ordering more torpedoes from Rains and would rejoin Hampton's command, but that never happened.[32]

Davis felt that the need for torpedo work in the Carolinas was so great that Rains should be sent. On February 21 orders were issued, dispatching him to Beauregard with sensitive primers and "some competent person as an assistant." As of March 7, Rains was at Greensboro, North Carolina, and Joseph E. Johnston, who had been named to command all forces opposing Sherman, was at Fayetteville. "I wish subterra shells in hands of my cavalry, now between Fayetteville and the Pedee, and beyond Goldsborough near Kinston," Johnston told Rains.[33]

The torpedo man was able to plant some landmines in fulfillment of Johnston's wish. Maj. Gen. John M. Schofield commanded a force of Federal troops to follow up the fall of Fort Fisher by advancing up both sides of the Cape Fear River and capturing Wilmington on February 22. Then he advanced toward Goldsboro with the intention of linking up with Sherman's columns. Schofield moved from New Bern by way of Kinston and was preceded by a detachment of the Twelfth New York Cavalry, which encountered landmines near Moseley Hall, about eight miles short of Goldsboro, in early March 1865. A horse touched one off, injuring the animal and the rider. A search followed during which more torpedoes were dug up. They were artillery shells "covered with a piece of tin and dirt" and with percussion fuses attached. In short, they were the handiwork of Rains, the same type of weapons he had planted at Richmond.[34]

Even though Wade Hampton had little faith in the ability of landmines to stop Sherman's progress, he admitted they could be useful in wrecking Union railroads. In late March he requested some Rains torpedoes along with "an efficient man" to plant them at the bridge near Kinston and along the track. But it was not possible at this stage of the war to organize such an effort.[35]

Rains continued to live in an overheated fantasy world. Writing to his daughter from somewhere in North Carolina on March 8, he assured her he had done "all possible to checkmate the Enemy." Rainy weather would slow

Sherman's march, and "if we are true to ourselves we will yet succeed."[36] Hampton's view of torpedoes was far more realistic than Rains's. A handful of landmines could not possibly retrieve Confederate fortunes, especially at this late stage of the war.

Sometime before April 3, Rains made it back to Richmond and became caught up in the evacuation of the capital. Burton N. Harrison, Davis's private secretary, had to deal with Rains and his family. The general wanted to bring along "what he considered a valuable collection of fuses and other explosives," Harrison recalled years later. "I distrusted such luggage as that, though the General confidently asserted the things were quite harmless." When Harrison told Rains there was no room in the cars reserved for Davis and his entourage, Rains went directly to Davis and inveigled his way into the party.[37]

After boarding, the train was delayed at the Richmond depot for some time. One of Rains's many daughters sat next to Davis. "That young lady was of a loquacity irrepressible, she discoursed her neighbor diligently—about the weather, and upon every other topic of common interest—asking him, too, a thousand trivial questions." Everyone else in the car felt depressed at the turn of events and remained silent. Only the voice of Rains's daughter could be heard. The only thing that lightened the moment was a sharp explosion that initially alarmed everyone in the car. One of Rains's staff officers had put a number of "torpedo appliances" in the pocket of his coattail and sat down too hard on a cold stove that stood in the car. One of them apparently went off but did no harm.[38]

Rains accompanied Davis for some time in the flight from Richmond, but calls for his devices continued to arrive. On April 7, Johnston once again sent a request for Rains to dispatch an officer "with material" to Wade Hampton. Even though the situation was hopeless, some Confederate authorities clung to the hope that landmines could play an effective role in averting their fate.[39]

It is true that the topography of South Carolina, with its numerous streams and causeways, represented a fertile field for the deployment of torpedoes as a harassment weapon. But the Confederate system of deployment was not efficient enough to try it on a large scale, and large scale deployment was the only hope to impose a significant delay on Sherman's and Schofield's movements. Only a handful of mines were deployed at Sister's Ferry and along the road to Goldsboro. They were quickly discovered and the mines were dug up, forcing minor casualties and delays of only a few hours. The Federals developed an effective and simple method of dealing with sub-terra shells in order to keep their operations moving forward with little trouble. Once again, while technically effective, Confederate torpedoes failed the acid test of tactical usefulness.

Chapter 9

Mobile

"A Thundering Report, a Flash, and the Groans of Wounded Men"

As the war continued to draw toward a close, Federal efforts to capture Mobile, Alabama, became the setting for the largest deployment of Confederate torpedoes that Union troops encountered during the conflict. Hundreds were laid out, mostly just before the Union army reached the zone of operations, and it is impossible to know with certainty how many were touched off or what the casualty rate might have been. Even the names of the operatives who planted them are obscure. But the episode left a big impression on Union soldiers, even though the torpedoes failed to alter their operations except to make them more careful about where they stepped. The plantings at Mobile ended the Civil War with a deadly flourish.

When Gabriel Rains suggested using torpedoes at Mobile to Dabney Herndon Maury in August 1863, as we have seen in chapter 6, the commander of the Department of the Gulf thought he was a crackpot. But by the early fall of 1864, Maury had been converted to their use. His engineer, Col. Samuel H. Lockett, placed fifty-four torpedoes by September 4 and continued with a large program until there were hundreds on land and in the waters of Mobile Bay by November 20.[1]

The torpedoes complemented a large system of defense constructed over several years to protect one of the more important blockade-running ports in the Confederacy. Mobile was located on the northwest sector of Mobile Bay that projected forty miles inland and was between ten and thirty miles wide. Ships entered the bay from the Gulf of Mexico through a narrow water passage protected by two antebellum masonry forts. The Confederates had also constructed earthworks to protect the land approaches to the city and planted earthworks at Spanish Fort and Blakely, two locations on the northeast sector of the bay.[2]

A Federal naval expedition had closed off the entrance to Mobile Bay by capturing the two coastal forts and defeating the small Confederate flotilla of naval vessels in the bay on August 5, 1864. Mobile was no longer a blockade-running port, but it was still in Confederate hands. Federal authorities wanted to use the city as a jumping-off place to project power into central Alabama, an area as yet largely untouched by Union troops.[3]

Maj. Gen. Edward R. S. Canby mounted a major effort to take the city in the spring of 1865. Gathering forty-five thousand men at New Orleans to oppose Maury's nine thousand troops, he transported his army by ship along the gulf coast. Canby landed thirty-two thousand men of the Thirteenth and Sixteenth Corps at the entrance of Mobile Bay with the intention of marching up the east side toward Spanish Fort. Maj. Gen. Frederick Steele, with thirteen thousand men, landed at Pensacola forty miles east of Canby with the intention of marching one hundred miles overland and approaching Blakely from the northeast. In the first of two landings by Canby's column, Thirteenth Corps troops debarked at Fort Morgan on the east side of the entrance to Mobile Bay and started to move north on March 17. In the second landing, Sixteenth Corps troops steamed six miles up Fish River, which emptied into the bay a short distance north of Fort Morgan, and commenced moving north by March 25.[4]

At the very start of the campaign, the Federals found torpedoes. A Sixteenth Corps brigade had landed at Cedar Point on the west side of the entrance to Mobile Bay with the intention of exploring the route north toward the city of Mobile, along the west side of the embayment. The Ninety-fifth Illinois led this brigade advance and soon spotted a number of landmines in the road. The regimental column barely stopped in time to avoid tangling with them. Clearing the road, the march resumed on March 19, and the next day, the Illinoisans found several more planted in the road. A decision was made to return to Cedar Point so the brigade could join the rest of Canby's column, riding on vessels across the bay to the mouth of Fish River on March 23.[5]

SPANISH FORT

From Fish River it was a short march of twenty miles to Spanish Fort for the two corps, but the way was paved with trouble. From early on the Federals began to encounter torpedoes. "We were greatly annoyed by those destructive devices from the time of leaving Fish River," wrote Col. William C. Holbrook of the Seventh Vermont. The roads "were strewn with loaded shell and covered over with a thin layer of earth." Engineer officer Lt. Charles J. Allen noted that they usually were planted in the ruts of the road to maximize the possibility of detonation by wagons and artillery pieces.[6]

Brig Gen. James I. Gilbert's Second Brigade, Second Division, Sixteenth Corps moved cautiously forward. On leaving Sibley Mills on Sibley Creek, the van of the brigade hit landmines. There was "a sudden crash like the bursting of a shell in our front," and members of the Thirty-second Iowa watched as Gilbert and his staff members who were riding at the head of the brigade became "enveloped in a cloud of dust and smoke." A horse had touched it off. The blast occurred just behind the horse ridden by Lt. Col. Jonathan Merriam of the 117th Illinois, stunning and throwing sand over him. Col. Risdon Moore of the 117th Illinois was only ten or twenty steps from the blast, which killed two horses but did not seriously injure any men. "The escapes seem miraculous," commented Monroe Joshua Miller of the regiment.[7]

Troops at the head of columns were the most exposed to these deadly devices, while the rest of the men became spectators of their effect. Virtually everyone could hear the explosions. They also saw the stacks of unexploded mines along the roadside after troops farther forward dug them up. The men of Gilbert's brigade counted eighteen torpedoes in a group as they passed the site of this incident, just north of Sibley Creek, on their way to Spanish Fort. "The Rebels here seem to be specially skilled in *torpedo* warfare," commented Monroe Joshua Miller. "This is truly an *enemy's* land. We are *'in perils every hour.'*" Rumors circulated that captured Confederates were forced to dig them up. The prisoners agreed with the Federals, according to these rumors, that it was not right to conduct war in this way, denouncing it as "base & cowardly." But more reliable reports indicated that the Federals themselves were responsible for clearing the roads. They learned how to discover the telltale signs of disturbance. "We soon found their mark where they were Put," as Louis Bir of the Ninety-third Indiana phrased it, and they dug them up without injury.

The Thirteenth Corps encountered a number of landmines as well. After pushing back mounted Confederate videttes, they began to touch off some explosions, "and every body is cautious," wrote Moses A. Cleveland of the Seventh Massachusetts Battery. Members of the Thirty-third Iowa found and took up fifteen torpedoes from the road on March 27 after a handful of horses were killed and some men wounded. As George Carrington of the Eleventh Illinois noted, most of the mines were discovered before they did their deadly work and were laid on the side of the road.[8]

D'Olive Creek, one and a half miles south of Spanish Fort, was the last natural barrier to Canby's march. As the Federals approached the stream, evidence of torpedo warfare increased. Lt. Col. John M. Wilson, assistant inspector general on Canby's staff, watched in amazement as a wagon approached the creek and hit a landmine. The blast destroyed the wagon and killed the mules and driver. Soon after that terrible explosion, Wilson decided to tear up the bridge the wagon was approaching because it was dilapidated, but he found a surprise on looking under the structure. He saw a "net-work of wires connected with a system of torpedoes arranged to blow up the bridge when a heavy weight like a piece of artillery should cross. Fortunately it had failed to do the intended work."[9]

The bridge over D'Olive Creek represented a combination booby trap and interdiction of troop movement toward the target and is the only example of a rigged bridging structure to be seen in the Civil War. It was, in other words, a more sophisticated example of torpedo warfare than any yet seen in the conflict, even though it failed in a technical sense. The numerous plantings of landmines along the roads, cunningly placed in the wheel ruts, also were more sophisticated than anything yet seen in the war.

We have little information as to exactly who planned and executed these new approaches to torpedo warfare in the Mobile campaign. Rebel soldiers recalled that men planted these mines for several days before the Federals made their appearance; that would seem to exclude Lockett, who had supervised mining operations several months before. There is no indication that he was responsible for this latest planting that took place just before the Federal arrival in March 1865. The only evidence as to who did it comes from an order issued by the headquarters of Brig. Gen. St. John R. Liddell, commander of the Eastern Division, Department of the Gulf. On March 31, Liddell's acting assistant adjutant general, H. L. D. Lewis, instructed the commander at Sibley Creek to retire the advanced pickets back to the creek in the face of Canby's advance. "You will direct the officer in charge of the squad with subterra shells to plant them at once, except in the road, where they will not be placed until the last moment, leaving a vidette in the road to inform any of our men coming in of the necessity of keeping in the middle of the road." A newspaper correspondent reported that two operatives were caught in the act of planting mines close to Spanish Fort and were compelled to identify the locations so they could be taken up.[10]

After crossing D'Olive Creek, Canby's men closed in on Spanish Fort, cutting it off by land. They were surprised to find even more torpedoes near the work than they had encountered on the road. Confederate operatives had planted them broadly and cleverly, not only just in front of the earthworks but scattered where Unionists were likely to walk. Federal troops found landmines on the approaches to springs and other watering holes. They were distributed through the woods and at a crossing of a small creek that flowed through the Union line. When the Ninety-sixth Ohio deployed as skirmishers upon initial contact with Spanish Fort, its members were surprised to find torpedoes planted on their side of numerous tree stumps, designed to blow up unwary Yankees trying to take shelter behind the stumps.[11]

Inevitably, this broadcast planting of torpedoes exacted a toll on the Union army. No one kept count, but reports circulated that a number of Federals fell victim to torpedoes. A specific incident occurred on the evening of April 5, when three men of Company A, Thirty-fifth Wisconsin touched off a landmine in the road next to a wagon. It took off both legs of one man, a leg and an arm of another, and the third also was badly injured. Lt. Col. James F. Drish of the 122nd Illinois related a story about another officer whose horse touched off a landmine. The steed was killed, but the officer was unhurt, even though he fell into the hole created by the torpedo immediately after the explosion.[12]

The Confederates saw torpedoes as an important element of their defense. Brig. Gen. Randall L. Gibson, who commanded the 2,200 defenders of Spanish Fort, was eager for all the landmines he could find. "Have received no more subterranean shells," he informed Liddell on March 27. "Many of those planted have bursted." By April 1 he strongly urged Maury to transport more mines to the fort and, two days later, repeated the message. "I can't get along without subterranean shells," he informed Liddell on April 7, adding that he desired "hand grenades—more negroes—a company of sappers & miners, a cutter or launch from the navy—two howitzers."[13]

From his headquarters in the city, Maury was well aware of the large-scale planting of landmines on the east side of Mobile Bay, but it appears that his chief engineer was not part of the process. Even though Samuel Lockett provided two detailed reports of engineering operations during the Mobile campaign of March to April 1865, he did not mention torpedoes in either of them. This probably means the engineering department had nothing to do with the landmines and a special corps of operatives was responsible for all the subterranean devices used in the campaign.[14]

When those operatives planted torpedoes near the works at Spanish Fort, they employed a carefully planned scheme. Federal observers noted that the landmines were placed between the ditch of the fortifications and an area of slashed trees fronting the works. The operatives planted them about six feet apart and in two lines so that the torpedoes of one line would cover the interval of the other line. This meant that there was a mine every three feet along the belt, maximizing the chances of an attacking column hitting them. Andrew F. Sperry of the Thirty-third Iowa called this a "diabolical plan of defense."[15]

Like Rains at Petersburg, these Mobile operatives planted small flags two inches square to mark the location of their mines. This was not only for the safety of friendly troops, but it would send a message to the Federals that an assault was unwise. A number of Union soldiers agreed that the torpedoes nullified the prospects of conducting a successful assault on Spanish Fort.[16]

Despite the gloomy outlook, Canby decided on April 8 that his men should launch an attack the next day at 8:00 a.m. Gibson also made a decision on April 8, but it was to abandon the fort after dark that evening. Federal skirmishers detected signs of the evacuation by midnight and began to carefully penetrate the torpedo field to investigate. They occupied the abandoned works in the dark and waited until daylight to look around.[17]

All were amazed at the number of landmines fronting the works and wondered that so few of their comrades were hurt by them. John S. Morgan of the Thirty-third Iowa heard that many of the torpedoes still had their protective caps over the fuse, although it is difficult to understand how the Rebel operatives could have neglected to remove them. Maybe the real reason for so few Union losses is that everyone was aware of the landmines and exercised extreme caution in dealing with the belt. When quartermaster Henry Fike walked into Spanish Fort to see the sights, he noticed "several torpedoes which had been found sticking their noses up out of the sand; but I took good care to not explode any of them."[18]

Nevertheless, some Federals fell victim to landmines after the fall of Spanish Fort. The perils were demonstrated by the story of an unidentified correspondent of the *Cincinnati Gazette*. He walked along the line of Rebel earthworks with Lt. George W. Fetterman of the Fifteenth US Infantry, who served as the commissary of musters on the staff of Brig. Gen. Kenner Garrard's Sixteenth Corps division, and with Capt. W. L. Scott of the Thirty-second Wisconsin, who served as acting assistant inspector general on the staff of Maj. Gen. Eugene A. Carr's Sixteenth Corps division. The three picked their way along and stepped up onto a traverse to get a good view. The correspondent and Scott made it safely, but when Fetterman began to step up, someone nearby yelled, "Take care, or you'll tramp on a couple of torpedoes there!" Fetterman instantly stopped and asked the soldier where they were. He said, "right down there," but the officer still could not see them. The man then coolly pointed directly at it with the bayonet on his rifle. Fetterman had placed his foot between two wires strung across the foot of the traverse; both of them connected to landmines. One wire was three inches from his leg and the other was six inches on the other side of it. He carefully lifted his leg away from danger and climbed up the traverse. "Myself and Captain Scott," wrote the correspondent, "who had walked in precisely the same path had probably set our feet still nearer the hidden death."[19]

This correspondent's story indicates that the Confederates had planted torpedoes inside their works at Spanish Fort. The traverse he mentioned would, of course, have been inside the fort, connected to the interior slope of the parapet for flank protection. These mines would have been planted at the last minute, just before Gibson's men evacuated the place. Maury's corps of torpedo operatives was active to the very last minute at Spanish Fort.

According to John S. Morgan of the Thirty-third Iowa, Federal pioneers began the work of clearing the mines from Spanish Fort on April 9, "taking up dead loads of them." John M. Wilson, Canby's inspector general, examined the devices and found that most were twelve-pounder shells arranged along the Rains pattern. A sensitive primer was inserted into a wooden plug with a tin cup placed over and touching the primer. When dug into the ground, the tin cup was an inch or two beneath the surface and covered with a layer of dirt. In places, a board was laid across several to increase the effect of an explosion by any one of them. Other men reported seeing twenty-four-pounder shells and asserted that the detonator was left above the ground. These variations in description might well indicate that the Confederate operatives did not plant the mines uniformly. Different operatives may well have done their job with slight variations.[20]

BLAKELY

As soon as Spanish Fort fell into Union hands, the focus of operations shifted to Blakely, five miles to the north. This fortified position was held by 3,500 men under Liddell and became the primary target of Steele's column marching from Pensacola. Steele arrived in the vicinity of Blakely by April 1. His cavalry captured a handful of Confederates who were engaged in planting torpedoes along the route of advance, one and half miles from the Confederate works. These operatives included W. R. Murphy and his comrades of Company C, First Mississippi Light Artillery. Liddell had "detailed us to plant torpedoes," Murphy later recalled, but it appears that the Union cavalrymen who captured them did not know they were deploying landmines. The prisoners were sent to Canby's headquarters and, from there, to Union-occupied Ship Island in the Gulf of Mexico for three weeks before transport to Vicksburg for parole and release. Because these Mississippi artillerymen were not forced to take up any of their own torpedoes, it is very likely they managed to hide the evidence of what they were doing just before capture and were therefore treated as regular prisoners.[21]

On April 3 Brig. Gen. James C. Veatch's First Division, Thirteenth Corps and Brig. Gen. Kenner Garrard's Second Division of the Sixteenth Corps marched from Spanish Fort to reinforce Steele. The reinforcements immediately encountered landmines along the way. When stopping to eat their noon meal near a stream, Veatch's men "struck a nest of torpedoes," as Moses A. Cleveland of the Seventh Massachusetts Battery put it. "Some exploded under wagon train killing mules and injuring drivers, casing much excitement." The column halted two hours as a result. On looking around, the Federals realized that the Rebel operatives had marked the location with blazes on nearby trees. Where there were no trees, they had "a stick put up in the shape of an X or cross." This helped the Federals to find and excavate up to forty mines. They piled them on the side of the road with a guard to make sure that no one bothered them.[22]

As they neared the works at Blakely, the Federals realized that their opponents had created a mine hazard similar to the one they had encountered at Spanish Fort. A slashing of timber was the first obstruction to face the Unionists. Then they found an abatis located fifty yards before the works and a line of inclined palisades with a wire entanglement four feet before the ditch. Operatives had planted torpedoes in a scattered form among these obstructions and quite near the ditch. For example, wires tied the palisades together for added strength, and additional wires ran from these strengthening wires to torpedoes planted nearby in the ground. Operatives also arrayed torpedoes in two lines fronting the works at Blakely, thereby creating an ordered formation to supplement the scattering of devices. They connected several torpedoes in these lines with wires to maximize the explosive effect on a wider area as all went off simultaneously. Small forked twigs subtly marked the location of at least some of the mines fronting Blakely.[23]

Because many of the torpedoes were scattered, the Federals at times dug their trenches among them without knowing it. The Thirty-seventh Illinois, part of Brig. Gen. Christopher C. Andrews's Thirteenth Corps division that reached the Federal force outside Blakely by April 9, reportedly dug its trench between two mined areas. Another regiment in the same division, the Ninety-seventh Illinois, dug its trench so near danger that a torpedo wound up just behind the trench, "on the rear edge" of it, in Andrews's words. Its location was not known until the regiment engaged in preparations for an assault ordered to take place on the evening of April 9. Capt. James W. Wisner of Company D was passing along the crowded trench when he touched off the device. It tore off his leg and badly injured two men, causing a delay in preparations for the attack. Wisner became the highest-ranking torpedo victim of the Civil War. As he and the other two men were carried back through the communication trench to the rear, hundreds of soldiers became visibly shaken by the bloody sight. "Such is the terror of concealed dangers," concluded Andrews.[24]

When the assault began on the evening of April 9, the Eighty-third Ohio led Col. Frederick W. Moore's Third Brigade in Andrews's division. "We went on scarcely firing, and not looking back," wrote Sgt. Thomas B. Marshall of Company K. The men knew full well that they were running through a torpedo field but did not let it hinder their movement. David Basset Snow mused that the Confederates "expected to see the air filled with mutilated Yankees. In this the villains were disappointed as but few of the torpedoes exploded when the advance was made."[25]

The other obstructions impeded the Federals more than did the torpedoes. The men had to use axes to cut paths through the slashing and abatis. Only when an individual happened to notice something suspicious did the torpedo have an effect on the troops. At one point the Confederates had stretched telegraph wire to trip an attacker, and the Federals assumed it was a wire connected to landmines. That caused some momentary hesitation.[26]

Of course when someone tripped a landmine, the results could be devastating. In the Ninety-seventh Illinois, Orderly Sgt. William R. Eddington took only a few steps when "a man next to me on my left stepped on a torpedo with his left foot. It blew his left leg off below the knee his right leg off above the knee and passed up between his head and mine, and never touched me. I grabbed him as he fell but I could not hold him."[27]

There were more torpedoes fronting Brig. Gen. John P. Hawkins's First Division of Steele's command, which consisted of three brigades of Black troops. Hawkins was positioned on the far right of the Union line fronting Blakely. When his men participated in the evening attack, they suffered more from landmines than did any other Federals on April 9. One torpedo alone severely injured six men in one company of the Fifty-first USCI. The regiment lost two who were killed and fifteen who were wounded due to all causes during the attack; one torpedo accounted for 35 percent of that loss.[28]

The Federals sustained a total of 127 killed and 527 wounded in their successful attack on Blakely, although it is impossible to know how many of those casualties were caused by torpedoes. No one collected data to find out,

Figure 9.1. Attack on Blakely. During the attack on the Confederate defenses at Blakely, a number of landmines went off, killing and injuring Federal soldiers. The explosion depicted on the bottom left of the image could represent a torpedo that affected the advance of the Federals. (Scott, *History of the 67th Regiment*, 94–95)

merely using vague phrases such as "several men" or "a great many" were lost. One soldier in the Sixty-ninth Indiana was reported as having had a leg blown off by a mine during the assault, and four men reportedly were killed by torpedoes in the Thirty-third Iowa, but there are no authoritative reports of total landmine casualties in Canby's force.[29]

Everyone agreed that even after the fall of the works, Federal soldiers became victims of the mines. Immediately after the charge, many regiments moved back to their camps in the Union lines and tangled with torpedoes on the way. The 114th Ohio marched in single file back through the obstructions until someone touched off a mine. Elias Moore of Company A was stunned by what happened. There was "a thundering report, a flash, and the groans of wounded men," he told his mother. "Soon I felt a shower of dirt falling over me. The report and flash were so unexpected and so near me that I was dazed for a moment, but was soon brought to my senses by the groans of three of our poor boys." Pvt. Harrison Teterick of Company A "was horribly mangled, one leg being blown off, the other broken and torn but still hanging to the body," wrote Moore. "Several pieces of the shell and metal had entered his bowels, and torn great holes in his body. His whole lower extremities were shattered and broken, his breast lacerated and his face horribly disfigured. He lived about two hours, was conscious most of the time and suffered very much. He called his near friends around him and sent messages to his parents. He was a brave soldier and a good boy. It was hard to see him die so."[30]

Teterick was not the only casualty of this torpedo. Sgt. Richard H. L. Walker of Company A "was severely wounded in the arm and back, only flesh wounds, but the flesh was almost torn from his right forearm," continued Moore. "Pvt. James Whitesides of Company A was badly wounded in the right thigh, the flesh being torn from the inside of the limb, also a wound in the arm. Both Walker and Whitesides survived their injuries. Another boy was badly stunned and powder burnt."[31]

The only loss in the Second Connecticut Light Battery, however, was John S. Mills, who stepped on a landmine while helping remove the wounded after the attack. The torpedo was not even near the fort but located in a road one hundred feet in front of the battery's position. Henry W. Hart saw the explosion and reported that Mills was horribly injured. His left leg was torn off, the right leg was wounded in two places, and his private parts were injured. Mills lived about three hours after that terrible moment.[32]

The landmines posed a major threat to the Federals as they tried to collect their wounded. By this time the sun had set, and there was no moonlight to help them. William R. Eddington of the Ninety-seventh Illinois received an order to go over the battlefield and make sure all the injured were found. He considered disobeying the order because he knew the ambulance corps could be relied on, and he also knew the extreme danger involved in walking between the lines in the dark. Fearing a court martial, Eddington steeled his nerves and went into the night. Instead of walking all over the field, he simply walked straight through the former no man's land to the field hospital located in a hollow. Along the way he expected "every step would be [his] last, but [he] got to the other side" of the most dangerous area. He talked to the attendants at the field hospital and then retrieved his knapsack to prove to the officer that he had actually gone back to camp, making his way by the same track he had earlier taken. "The same kind providence that shielded me in so many close calls was still with me," he wrote.[33]

Pvt. Josias Lewis of Company K, Forty-seventh USCI was not so lucky. He was assigned to escort prisoners to the rear soon after the attack and stepped on a landmine along the way. Col. Hiram Scofield, his brigade commander, personally witnessed the explosion. Lewis lost a leg as a result. Andrews reported that some of the Confederate prisoners taken to the rear tried to lead the way safely through the mines to save themselves, but that also saved their guards.[34]

Many men had a natural curiosity to see the Confederate position and braved the threat of destruction by torpedoes. Henry W. Hart of the Second Connecticut Battery counted fifteen torpedoes within a space about fifty feet in diameter. "While counting, I found myself standing astride one, my right foot being about 3 inches from the cap. I walked out of it very careful as though I was walking on eggs," he told his wife.[35]

Other curiosity seekers included Lt. John M. Godman and Commissary Sergeant Henry S. Beacher of the Ninety-sixth Ohio. When a sentry warned them of the landmines, they continued "with that listless indifference to danger that so mysteriously becomes part of a soldier's being." They could see in the dim twilight that many torpedoes had already been marked with stakes and guards were placed to warn stretcher bearers carrying the wounded through the danger zone. The pair even saw holes created by the explosion of torpedoes. Godman and Beacher made it to the works and spent so much time looking around that it was very dark when they started back to camp. The guards were gone, and they could not see the stakes. Stepping carefully, helped a little by those men with lanterns who were still searching for wounded, they made it safely out of danger.[36]

Other Federals apparently were not so lucky. Several accounts indicate that the sound of exploding torpedoes could be heard off and on during the night of April 9 as stretcher bearers and curiosity seekers continued to walk about in the dark. Henry W. Hart counted about fifty such explosions during the night.[37]

Curiosity seekers also began an unofficial clearance of the mine field in their own fashion. Wondering what these devices looked like, a handful of Unionists dug them up to see. Members of the Ninety-seventh Illinois managed to excavate one without exploding it, and W. R. Mallory of the Seventy-second Ohio personally dug one up.[38]

Impromptu digging could not be relied on to clear the torpedoes; only orders from on high and the employment of many men could take care of this dangerous job. Federal officers were keen to use the hundreds of Confederate prisoners captured at Blakely for that purpose, but they found some difficulty in that regard. Many of the captured Confederates were afraid they would suffer because of the torpedoes planted during the campaign. When the Eighty-third Ohio charged into the works at Blakely, some members cried out, "Torpedo," to remind their comrades of the danger. The Confederates heard this and interpreted it as a cry of revenge. "They threw down their guns and begged for mercy, for they thought they would be killed because they had planted the ground full of 'torpedoes,'" wrote Isaac Jackson. The Rebels begged their captors to be treated as prisoners of war. "They knew they had of [sic] done wrong by fighting with torpedoes."[39]

David Basset Snow of the Eighty-third Ohio found about a dozen Confederates huddling in a bombproof inside the works. He drew his rifle on them and demanded they give up. "One by one they came out asking for mercy & trembling like frightened children. 'We all didn't put down the torpedoes,'" they said to Snow. "I told them to 'limber to the rear & keep mute.'"[40]

Some Federals were angry about the landmines and felt the urge to exact some sort of revenge on their prisoners. Maj. Gen. Gordon Granger, commander of the Thirteenth Corps, expressed his feelings upon seeing prisoners being led to the rear soon after the assault. According to Moses A. Cleveland of the Seventh Massachusetts Battery, temporarily detailed to escort captives to the rear, Granger shouted, "March them over the Torpedoes, ___ them." But no revenge was exacted on the captives.[41]

While many enlisted Confederates shuddered at the thought of retaliation, at least some officers maintained a defiant attitude. Col. Elijah Gates refused to cooperate with the Federals when they asked him to locate the mines. "I don't know where they are, and by the lord Harry if I did I would not tell you." They threatened to make him walk about where Federals were walking, but that did not convince the colonel to change his mind. "All right, I have the satisfaction of knowing you will be blown up with me."[42]

But many other captives were willing to help the Federals in mine clearance, or to put it another way, they were willing to obey when their captors told them to help. Frederick Steele issued orders to retain enough prisoners at Blakely for that purpose while sending the rest to Spanish Fort. Details of prisoners were busy at the task around the works at Blakely on April 10. They were guarded by Black troops; "a good employment for *both*," thought Moses A. Cleveland, who admired the bravery shown by Hawkins's men in the assault the previous day. When Henry W. Hart of the Second Connecticut Battery heard that two prisoners had been killed by the torpedoes, he thought it was an appropriate fate for them.[43]

There is no confirmation of casualties during the cleanup process and no clear proof that the Federals employed coercion in using prisoners for that purpose. Rumors insisted that the Federals threatened to march captives in a mass over suspected ground before the prisoners agreed to participate in the cleanup. Joseph T. Woods of the Ninety-sixth Ohio heard that a captain who had been part of the torpedo-planting squad was among the captives, but he refused to locate the mines. The Federals put him in a group of one hundred prisoners, formed in a column eight men wide, and made the formation march along roads not yet cleared. Guards at the rear had orders to shoot down anyone who hesitated. The captain finally agreed to help clear 185 torpedoes from a quarter mile stretch of the road, according to this story.[44]

Whether rumors such as these can be credited is questionable. But we know for sure the prisoners played a huge role in mine clearance on April 10. Thomas B. Marshall of the Eighty-third Ohio saw many squads of them at work and counted eleven torpedoes in the narrow sector through which his regiment had charged the evening before. Details were in the process of taking the devices away and detonating them in a safe location. The process of safely detonating them was crude but apparently effective. The men placed a burning pine knot on them according to one report, while another had it that the prisoners spread brush over a wide space of ground and set it afire, hoping to set off any torpedoes that happened to be within the brush area. In some places rain had washed away enough dirt so that the caps of the mines were visible to the naked eye, helping greatly in the task of locating them.[45]

While taking inventory of material left by the Confederates inside the works, the Federals discovered at least twenty torpedoes that had not been planted. The number may have been higher. Moses A. Cleveland saw the landing on the Tensas River near Blakely on April 14 and noted that Federal quartermasters were shifting material captured five days before to it. "On the wharf are guns, torpedoes, and boxes of muskets." These mines probably were those that had not been planted, for there is no indication the Federals defused torpedoes they had taken out of the ground. They all seem to have been destroyed rather than retained.[46]

Observers who commented on the widespread use of torpedoes at Mobile agreed that the physical damage to the men was limited but the emotional effect was much greater. "The torpedo was an unaccustomed weapon of warfare to us," wrote Andrew F. Sperry of the Thirty-third Iowa. His comrades tended to be more nervous about them than of bullet and shell and "would examine the ground minutely for the little sticks which served to mark the place where a torpedo was buried."[47]

It will come as no surprise, then, that every Union soldier who commented on the Confederate use of torpedoes at Mobile was angry or critical. "This kind of warfare is nothing short of cold blooded murder," Elias Moore told his mother on April 10. "I would like to see the fiend who ordered them put in the ground, hung, shot, or burnt. There is no death too severe, no torture sufficient to retaliate for the murder of our brave boys." Monroe Joshua Miller of the 117th Illinois was equally incensed. "The most inhuman savages of heathendom ought to be ashamed to resort to such mode of warfare as this torpedo business." Others expressed their feelings with more restraint. Isaac Jackson of the Eighty-third Ohio called torpedo warfare "a very mean way of fighting," while Henry W. Hart of the Second Connecticut Battery wrote, "I think it is not civilized."[48]

Many Confederates sensed that the scale of torpedo planting at Mobile was something unusual and that it would spark a strong desire for retaliation among the Federals. They were right. For Lt. Thomas L. Evans of the Ninety-sixth Ohio, these mines were a turning point in his attitude toward the enemy. "Up to this time, I have felt like dealing honorably even with Rebels but the last spark of such feelings is gone. The whole army here now feels like nothing could be unjust in dealing with such unscrupulous traitors. I cannot express my feelings toward them, it is such as I have never felt before."[49]

Rumors once again came into play to express the spirit of retaliation against the Confederates for their use of landmines. Lt. Col. Oran Perry of the Sixty-ninth Indiana was wounded in the assault of April 9 and further injured by a torpedo when making his way to the rear. Elias Moore heard that it was "only with difficulty his men could be restrained from retaliating on the prisoners." Another report indicated that Canby wanted to use Greek fire to bombard Spanish Fort and Blakely in retaliation for the torpedoes his men encountered on their approach to the works. Five thousand shells containing the chemical composition that made up Greek fire reportedly had been unloaded from transports by April 8, but then Gibson evacuated Spanish Fort that night and Blakely was captured the next day before they were used.[50]

In the final analysis, there is no solid proof that the Federals exacted any revenge on the Confederates for their use of torpedoes at Mobile, except if one can count the use of prisoners of war in that regard. Anger fueled the desire for retaliation, but the act did not follow.

Emotions aside, when the Federals came to evaluate the tactical effectiveness of the Mobile landmines, the type of detonation system became all-important. Engineer Lt. Charles J. Allen noted that the torpedoes at Spanish Fort were all touched off by contact, which meant that only one man could detonate them. "When we consider the small space taken up by a 12 pdr. shell, and the distances by which they were separated, it is not to be wondered at that they produced so little devastation." Allen described the detonation device as "a metallic plug similar in shape to the common fuze-plug, through which passed a needle, the latter falling upon a nipple with the ordinary percussion cap."[51]

But Allen failed to note that some landmines at Mobile were wired. Friction primers in wooden sleeves that were fitted into the fuse adapter and with a wire attached to the end set them off. As we have seen, several mines were linked by wire so that any disturbance of the system should have set off all or most of the individual mines. But there is no evidence that these linked systems worked better than individually installed mines. Federal accounts of torpedo injuries failed to differentiate between the system and the individual mine.[52]

John M. Wilson, Canby's assistant inspector general, asserted that yet a third detonation method was used at Mobile. In addition to Rains's sensitive fuse, the Confederates deployed some torpedoes with a chemical reaction device. "A small glass tube of sulphuric acid was contained in and rested against the head of a soft cap; surrounding the tube and holding it in position was a mixture of chlorate of potassa and white sugar, under which was a quick-burning composition; upon striking any rigid body, the soft cap was crushed, the glass tube broken, the sulphuric acid in contact with the chlorate produced fire, which was communicated to the charge through the quick composition." Someone at Mobile must have read Richard Delafield's report of the Crimean War operations and duplicated Immanuel Nobel's device, which was used by the Russians at Sebastopol. While rumors indicated that

Figure 9.2. Ten-Inch Sub-Terra Found at Blakely. This ten-inch mortar shell was among hundreds planted by the Confederates to aid in the defense of their fortifications at Blakely. (AHC and Herbert M. Schiller, *Confederate Torpedoes*)

a British officer named Green had taught the Confederates how to make and plant landmines, there is no evidence to support those rumors.[53]

It is not possible to arrive at the number of men killed or wounded by torpedoes at Mobile, because no one filed a report on the matter. While some men believed torpedoes took down more of their comrades than Confederate rifle or artillery fire, that is extremely unlikely. Many other Federals believed torpedo losses were low compared to the number of mines they found. Canby's total losses during the campaign for Mobile amounted to 1,508 men (177 killed, 1,295 wounded, and 36 missing). Of that number, probably between 50 and 100 were due to landmines.[54]

The Federals believed that at least one of their enemies had been killed because of torpedoes. After occupying Spanish Fort, they found the body of a Confederate in front of the work who "appeared to have been killed by a musket-ball while planting a torpedo," wrote division commander Andrews. "Close by him was a spade; also a torpedo planted about two feet in the ground, but not yet covered. They buried him there with it."[55]

Those torpedo victims who were wounded rather than killed wound up in field hospitals, where their terrible injuries became common knowledge. Sgt. William J. Gould of the Second Connecticut Battery visited the hospitals, where he found that "the worst wounds were by Torpedoes."[56]

Animals probably were affected more by these landmines than men. Not only the cavalry units attached to Canby's command but infantry units lost horses and mules. Their large bodies tended to absorb much of the blast. As a result soldiers mounted on horses and mules often wound up with wounds rather than death. "Quite a number of horses were killed, and several men wounded, some mortally," was the way William H. H. Clayton of the Nineteenth Iowa put it.[57]

The total number of mines planted during the Mobile campaign also is difficult to determine. The Confederates maintained their torpedo program as a semisecret weapon, and we have no official reports or unofficial accounts by the men who actually deployed the mines. The few references to torpedoes in the available Confederate records have already been cited earlier in this chapter. Unfortunately, Canby did not assign anyone to accumulate data on the number of landmines dug up after the campaign. The same is true of the submarine mines planted in Mobile Bay, which were a major problem for the naval officers who tried to support Canby's land operations.[58]

But there are some estimates of the number of landmines used in the campaign, although they are not backed up by firm documentation. Historian William C. Schneck asserted that 205 Rains torpedoes had been planted at Spanish Fort but did not indicate where he found that information. Two Federal soldiers made offhand reference to numbers in their personal accounts. Henry W. Hart of the Second Connecticut Battery mentioned that

1,000 landmines had been planted "around here," presumably referring to Blakely, while David Basset Snow of the Eighty-third Ohio referred to 900 torpedoes planted at Blakely.⁵⁹

We can safely conclude that the Confederates deployed about 1,000 to 1,250 landmines at Mobile. That is roughly in the same range as the torpedo belt at Petersburg, where at least 1,298 and as many as 2,363 mines had been placed. But the planting at Mobile was more diverse, widespread, and deadly than at Petersburg. Maury's torpedo men not only placed them before earthworks but spread them out over great areas, covering roads, paths, and stream crossings. They mixed in some booby traps as well. The operatives also used three different types of detonation—mines that were individually set off by pressure from above; mines arranged in a series to be set off by the tug of wires; and a few chemical reaction detonators, which also were set off by pressure from above. The only detonation device not used at Mobile was electricity.

Yet, once again, the extensive work of these torpedo experts failed to halt the Federals. It failed even to slow them down appreciably. All it did was add yet another worry to their operations, a few dozen additional casualties, and a lot of anger and bitterness toward the Confederates. Rains's fervent hopes that the new mode of warfare he pioneered would have immediate and large-scale effects on enemy operations were never realized during the Civil War.

Chapter 10

Attacking Communications

"He Did Not Want the Matter to Become Public"

Several Federals and Confederates worked to improvise explosive devices designed to interrupt their opponents' means of transportation and communications. Some of these devices bordered on ingenious in their design and use, lending themselves to covert operations behind enemy lines. As with landmines, these explosives worked technically, played a role in the tactical thinking of both sides, and raised more moral arguments, but they did not pay off in terms of seriously hurting the intended targets except in limited ways for a handful of incidents.

The Union army caved in to the developing strains of torpedo warfare more than halfway through the conflict, but it focused entirely on weapons designed to disrupt travel by the Confederates rather than antipersonnel devices. Soon after the fall of Plymouth, North Carolina, to a Confederate force on April 20, 1864, Brig. Gen. Innis Palmer ordered Lt. W. R. King "to construct some Torpedoes" to plant in the sounds along the coast of that state. Palmer, commander of the District of North Carolina, wanted to interrupt the movements of Confederate ironclads, which now had the opportunity to steam down the rivers to the coast. King put together a number of underwater mines and, by late May, began to move them to the Roanoke River to block the movement of *CSS Albemarle*.[1]

A terrible accident took place during the movement of these torpedoes. A shipment of them arrived at Batchelder's Creek, nine miles west of Union-occupied New Bern. The creek was the forward zone of Federal control in this region of North Carolina, with a strong picket line along the east bank. The Federals operated the Atlantic and North Carolina Railroad from New Bern out to their post at the creek. On May 26 a train bearing the last four mines of a shipment of thirteen to be placed in the Neuse River arrived at 4:00 p.m. Each contained 250 pounds of gunpowder in barrels wrapped with iron bands. While unloading the last of the four, a log somehow hit the cap and exploded it. That explosion set off the other three, which were resting on the platform.[2]

The resulting blast, reportedly heard thirty miles away, devastated the crowd near the center of the explosion. "Heads, bodies and limbs were scattered for a quarter of a mile around," read one report of the scene. "Many are torn to atoms," wrote Lt. Cornelius C. Cusick of the 132nd New York, the regiment most affected by the blast. Bodies were so mutilated that friends could not recognize them. Regimental commissary sergeant David Jones could only be identified by a ring on his dismembered arm. Survivors quickly filled three hardtack boxes with "fragments of flesh picked up on the spot." The blast sprayed blood and human flesh all over the locomotive and battered it in places while partly demolishing a passenger car. A commissary building and a signal tower were destroyed.[3]

A total of thirty-one men were killed and twenty-two wounded in the Batchelder's Creek explosion. Of the fifty-three casualties, forty-two served in the 132nd New York and five in the 158th New York. Four more soldiers served in other regiments. Two civilians, an eleven-year-old boy who was injured, and the Black servant of a Union officer who was killed were listed among the casualties. The boy, named Frank Gould, pleaded with army surgeons. "Doctor, I can stand any amount of pain, but don't take off my leg.'" At last report, the surgeons agreed with him.[4]

The first news of this "fearful explosion" was sent by Col. Peter J. Claassen of the 132nd New York. "Send coffins," he asked of the garrison at New Bern. Writing only a half hour after the accident, he could not tell much about the cause except an assumption of "idle curiosity" on the part of onlookers. "I am too sick at heart to tell you more just now," he concluded.[5]

Palmer was very concerned about the cause of the accident. He heard that the torpedoes had been hauled to the train at New Bern in an open wagon with no supervision, that they had been dumped as if they were "the same as barrels of provisions" at the depot, and that they were unloaded at Batchelder's Creek by Black laborers with no one to superintend the work. The general did not credit this report, because he had too much faith in King, but he initiated an inquiry to determine exactly why the four mines had exploded.[6]

There are no documents indicating the result of Palmer's inquiry, which leads one to assume that King was not culpable of neglect. The fact that he continued to direct the Union mine program in North Carolina is another indication that Palmer was satisfied with his work and that careless handling by the Black laborers (who were not listed among the dead) was the cause of the explosion. Two more unlisted victims of the blast were Pvt. Henry B. Tibbetts and Pvt. Amos P. Barnes, detailed to the US Signal Corps. They were killed when the signal tower at Batchelder's Creek crashed as a result of the blast.[7]

King continued with his torpedo program after the terrible accident, planting mines in the Roanoke River that he believed prevented *CSS Albemarle* from descending the stream. Federal gunboats had to act to prevent the Confederates from removing the mines. About August 1, 1864, Maj. Gen. Benjamin Butler approved a suggestion that his engineers place a number of submarine mines at the mouth of the Cape Fear River to prevent blockade runners from using Wilmington as a port of entry. But a month later, when the idea for an expedition to capture Fort Fisher came forth, this idea was dropped.[8]

The Federals expanded their torpedo plans to include devices designed to disrupt railroad traffic. Herman Haupt was responsible for this development. A civilian railroad expert hired by the government for war service, he also was particularly interested in bridges. On November 1, 1862, Haupt wrote his thoughts about using explosive devices to bring them down. Noting that most railroad bridges in Virginia consisted of "Howe trusses without arches," he thought they could easily be wrecked by breaking one end of the bridge or, at least, one section of it. Two torpedoes could be placed in one section in two separate holes bored with a hand auger. An arch would require at least four torpedoes.[9]

Haupt conceptualized a plan to employ these devices to suit the tactical requirements of the Federals. He theorized that the devices had to be quickly planted in less than five minutes, that the equipment had to be minimal in number, and that each item had to be small enough to be carried in a pocket or a saddle bag. The system would be useful during cavalry raids deep behind enemy lines or for infantry forces compelled to give up territory and trying to disable transportation lines before they left the region.[10]

The Haupt bridge torpedo was designed to meet these tactical requirements. It consisted of "a short bolt of seven-eighths inch iron, eight inches long, with head and nut, the head to be two inches in diameter and about one inch thick; a washer of same size as the head must be placed under the nut at the other end, with a fuse hole in it; between the washer and the head is a tin cylinder, one and three-quarters inches in diameter, open at both ends, which is filled with powder, and when the washer and nut are put on, forms a case which incloses it." A hand auger to drill holes and a proper fuse were also needed. Once the hole was drilled, Haupt instructed operatives to push the torpedo in headfirst and secure it by tapping with a stone or piece of wood. He recommended "ordinary cigar lighters, which burn without flame and cannot be blown out," for igniting the fuse.[11]

Ever the professional, Haupt conducted experiments to see if his device would work. He blew up tree trunks with it and simulated a bridge truss at the

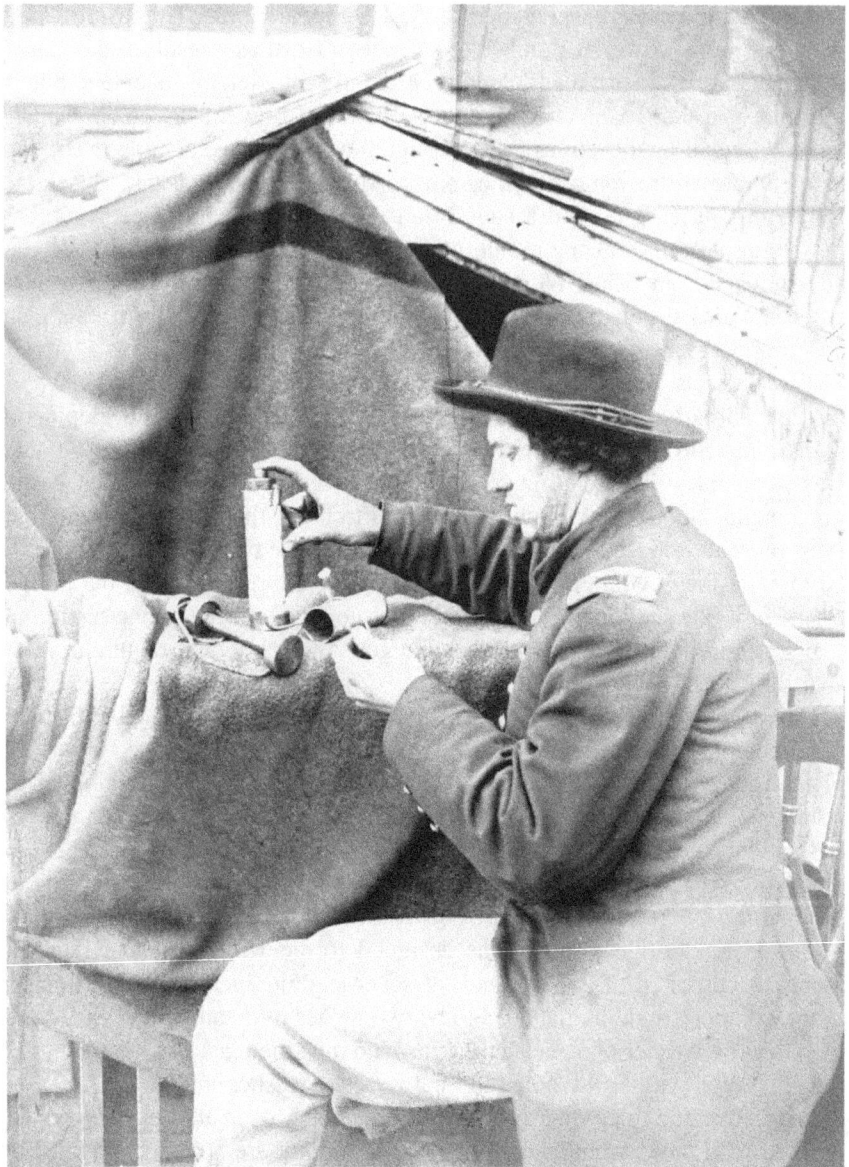

Figure 10.1. Bridge Torpedo. A Federal officer examines the components of a torpedo developed by railroad engineer Herman Haupt for quick and easy destruction of bridges. It is an Andrew J. Russell photograph. (LC-DIG-ppmsca-10403)

US Military Railroad depot at Alexandria to demolish it. Everything worked as planned, and Haupt instructed his subordinates to provide samples of the torpedo to any officer requesting information.[12]

Haupt then readied his device for field use. As Burnside was planning his move across the Rappahannock River at Fredericksburg in early December 1862, Haupt arranged for one hundred bridge torpedoes to be made, but he wanted to paste a diagram on each one to illustrate a bridge truss, along with written instructions about how to insert the device. This illustrated instruction was to be lithographed in Washington, DC. Haupt's subordinates packed the torpedoes in boxes, fourteen to a box; they were ready for use, loaded and with fuses inserted. Haupt instructed them to pack extra copies of the instructions, extra fuses, friction fuse lighters, augers, and handles in every box so that all would be "complete for use."[13]

Assuming Burnside could cross the Rappahannock, the idea was that the Federals would advance south along the Richmond, Fredericksburg, and Potomac Railroad. Cavalry forces would range far ahead and destroy bridges behind Lee's Army of Northern Virginia to disrupt its supply line and hamper its farther retreat south. But that never happened, because the Federals were bloodily repulsed when they attacked Lee's position on the south side of the river on December 13.

There are only two other references to Haupt's bridge torpedoes that suggest they were actually used, or intended to be used, in the field. During the Chancellorsville campaign, Maj. Gen. George Stoneman conducted a cavalry raid to destroy rail connections between Fredericksburg and Richmond. A bridge that carried the Richmond, Fredericksburg, and Potomac Railroad across the North Anna River became a target of the First US Cavalry on May 2, 1863. It consisted of two posts in the center, according to trooper William Fawcett. As skirmishers deployed to keep opposing Rebels busy, an officer told Fawcett to bore a hole in the main post and place a bridge torpedo. He did so, lit the fuse, and retired to watch the result. According to Fawcett, the blast only partially destroyed the structure. "The bridge swung down out of shape, but the other post held it up." The Federals then used an age-old technique, setting the wooden structure on fire. Despite this partial failure, Stoneman's chief quartermaster suggested after the raid that every such expedition should "be provided with a number of torpedoes, with proper sized augers" for destruction of bridges.[14]

The only other reference to Haupt's bridge torpedo appeared a year later, in the early phase of the Petersburg campaign. Grant authorized a cavalry raid by Brig. Gen. James H. Wilson and Brig. Gen. August V. Kautz to disrupt Lee's railroad supply line approaching the town from the west. Montgomery C. Meigs shipped five hundred sets of portable rail twisters developed by Haupt and requested bridge torpedoes of the Ordnance Department, indicating that, by this time, the army had taken responsibility for making the device. Neither the twisters nor the torpedoes arrived in time before the expedition set

Figure 10.2. Applying a Bridge Torpedo. An operative demonstrates how to drill a hole in the supporting timber of a bridge for insertion of Haupt's bridge torpedo. An officer sitting on the right is holding the device. This is another Andrew J. Russell photograph. (LC-ppmsca-10406)

out on June 22, 1864. The men improvised methods of destroying track and resorted to burning bridges.[15]

The sparse evidence that Haupt's bridge torpedo was ever used in the field lacks a clear explanation. It is true that Federal forces only sporadically were in a position to need it, but one wonders if the general disdain for torpedoes among most Federals might have played a role in this as well. If Fawcett is to be trusted (and his postwar account tends to be short on details), the device either failed to live up to expectations, or Fawcett erred in placing a torpedo only in one post instead of both. Despite Haupt's apparently effective design, his thorough testing, and his promotional efforts, his bridge torpedo played only a minor role in the war.

Another effort by the Federals to disrupt enemy transportation appeared in the fall of 1864, when Capt. Charles F. Smith of the Third US Colored Infantry devised what appeared to be an effective railroad torpedo. Serving in the Department of the South, Smith suggested his superiors send small parties of three or four men behind enemy lines at night and plant the devices on Confederate railroads. "Each magazine is a load for a man," wrote Lt. Charles R. Suter, chief engineer of the Department of the South. "Another man can carry the lock and another the tools." Maj. Gen. John G. Foster, commander of the department, ordered Suter to send a sample torpedo, the tools, and drawings demonstrating its use to the chief engineer of the army, Richard Delafield, in Washington, DC.[16]

Suter also wrote a detailed explanation for Delafield. The method of planting consisted of pulling the spikes of three ties that lay next to each other. Then the operatives were to dig a hole for the torpedo on one side of the three ties and spring the rail so that it raised a short distance above the ties. They then were to place what Suter called a "lock," but probably more accurately termed a bar of some kind, so that one end rested between the raised rail and the ties. After placing the charge in the hole, the other end of the lock was to be attached to the detonating device. Iron wedges between the raised rail and the ties prevented the lock from tripping the charge of its own weight, but the enormous weight of a locomotive would force the rail down enough to explode the charge.[17]

Suter tested the device with eighteen pounds of gunpowder under a spare railroad car. The vehicle was "entirely destroyed, and rails, ties, and fragments of the car were thrown in every direction. One rail was projected 40 feet." He thought the best part of the device was that it would destroy the locomotive and cars in addition to merely disrupting the track. Encouraged by the trial, Suter suggested loading the charge with twenty to thirty pounds of powder for greater effect. He believed one lock would set off two charges at the same time and "by regulating the length of the powder train, any car of the passing train may be blown up." He also suggested applying tar coating to the magazine that held the powder to waterproof it.[18]

The development and testing of Smith's torpedo took place before Foster's troops mounted a couple of raids against the two railroads that crossed northern Florida in late July 1864. Smith was detached from the Third US Colored Infantry and ordered "to select a company of men and to drill them in the use of the railway torpedoes," but he failed to do so and was sent back to duty with his regiment. Suter assigned Capt. George Edwards and Company L, First New York Engineers to the job.[19]

Brig. Gen. William Birney commanded the first raid, starting from Jacksonville, Florida, on July 15, 1864, and landing at the mouth of Trout Creek, eight miles north of the city up the St. John's River. He took along four torpedoes and managed to move part of his force as far as Callahan on the Florida Central Railroad, twenty miles northwest of Jacksonville. But the Federals did not use Smith's torpedo. For unexplained reasons, Birney ordered one of them to be destroyed and took the other three back to Jacksonville. Brig. Gen. John P. Hatch led the second raid, leaving Jacksonville on July 23 and hitting the Florida Central Railroad at Baldwin, the point where that line intersected with the Florida, Atlantic, and Gulf Central Railroad. The Federals burned several trestles along both lines the next day, before returning to Jacksonville. Hatch had taken along one of Smith's torpedoes but did not use it. In fact, like Birney, he destroyed it in the field for unstated reasons. Following these two raids, twenty more Smith torpedoes were constructed and remained in storage.[20]

That is why Suter complained to Delafield on October 26, 1864, that "we have not as yet been able to try them on the enemy's railroads." And they would never be tested. Birney relied on tried-and-true methods of rail destruction: burning bridges and tearing up track. He could have planted Smith's device to harass the Confederates when they came to repair the damage but failed to do so. Moreover, why he and Hatch felt the need to destroy two of them is inexplicable.[21]

Smith came up with what appeared to be an effective railroad torpedo, even though it involved a complicated process of planting. In fact, only a trained operative could have set it up properly. That complication may have played a role in the fact that it was never used in the field. Smith himself failed to provide the expertise or promotional effort to popularize the device. He had initially enlisted as a private in the Tenth New York Heavy Artillery and was promoted to captain of Company I, Third US Colored Infantry on September 6, 1863, at the advanced age of fifty-two. He brought into service a long problem with alcohol but swore off drinking for two years while serving in the army. Unfortunately, being detached for duty at Hilton Head led him to fall off the temperance wagon. Smith "has relapsed into excessive drinking," reported Lt. Col. Ulysses Doubleday of the Third US Colored Infantry. "He has a large family, & is an excellent officer, too good to go to utter ruin." Doubleday requested that Smith return to the regiment to escape temptation. But just before that transfer, Suter asked Smith to train the operatives for the Birney-Hatch raids. Probably still feeling the effects of his drinking spree, Smith failed to do so and returned to his regiment. He was allowed to honorably resign at the end of the war. Like Haupt's bridge torpedoes, Smith's railroad torpedoes failed to receive a proper field test.[22]

Federal forces made an attempt to use a torpedo to destroy a railroad locomotive along the Richmond-Petersburg line in August 1864. The Confederates had been using a portable steam engine to operate a mill behind their picket line near Dutch Gap. Capt. Edgar A. Nickels of Company C, Eleventh Maine led an expedition consisting of three companies of that regiment and a detachment of forty men from the Third Pennsylvania Heavy Artillery to destroy the engine. They obtained a torpedo from the US Navy and set out up the James River by boat, landing near the picket lines on the north side of the stream early on August 3, 1864. As he neared the target, Nickels deployed one company as skirmishers and formed the other two in column behind the right and left wing of the skirmish line. The Pennsylvania heavy artillery men remained behind the center of the line to protect the operatives who carried the torpedo.[23]

Fifty yards short of the mill, Nickels encountered Confederate skirmishers and pushed them back two hundred yards to obtain complete possession of the abandoned mill. But despite a thorough search, the Federals could not find the engine. They detected signs that it had just recently been removed, so they tried to use the torpedo to destroy other equipment left behind. Operatives placed it next to a couple of large iron wheels and shafts and attached a lanyard, but the device would not cooperate. They tried to fire it three times, but each time, the friction primer failed. There was no use trying anymore, so the expedition retired.[24]

Nickels's effort to destroy the Confederate mill came to naught for many of the same reasons that Federal efforts to disrupt enemy communications failed. The Federals were only rarely presented with the opportunity to plant mines on railroads, and technical limitations hampered the effort even when they tried to do some damage. The naval torpedo failed miserably, the Smith torpedo's delicate placement demanded an expert to use it, and Haupt's bridge torpedo was inexpertly deployed by Fawcett.

When the Federals wanted to destroy Rebel railroads, they more often relied on cavalry raids to burn structures and rolling stock, as had Birney in his north Florida raid. Or, as Sherman preferred, they used large infantry forces to occupy those lines for extended periods of time and thoroughly wreck the roadbed so that it would take months for the Confederates to affect repairs. This is what Sherman did in the last stage of the Atlanta campaign, supplementing the action with booby traps. Moving six corps of his army group from their own line of supply and marching them to points southwest of Atlanta, Sherman cut the last two railroads feeding John Bell Hood's Army of Tennessee in the city. On August 29, 1864, his men tore up track along more than twelve miles of the Atlanta and West Point Railroad, filling in cuts with "the trunks of trees, logs, rocks, and earth, intermingled with loaded shells prepared as torpedoes to explode in case of an attempt to clear them out." Sherman reasoned in his memoirs that even if only one shell went off, it "would have demoralized a gang of negroes and thus would have prevented even the attempt to clear the road."[25]

After Sherman left the region and the Confederates were ready to restore the line, Beauregard wanted to use Federal prisoners of war to clear out the cuts along the Atlanta and West Point Railroad. He requested Brig. Gen. John H. Winder, the Confederate commissary of prisoners, to furnish thirty Federals to Maj. James M. Hottle, the quartermaster in charge of the railroad work. Because Beauregard's assistant adjutant general told Winder that the Federals would be used to clear out the torpedoes from the cut, Winder refused to comply. "I don't think this is legitimate work for prisoners of war," he wrote while passing the request on to the Adjutant and Inspector General's Office in Richmond. That office sent the request to the Confederate War Department, where Assistant Secretary of War John A. Campbell agreed with Winder unless the prisoners were "willing to do so." This reply reached Beauregard in mid-February 1865, a month after his initial request. Still, nearly another month after that, Beauregard responded to the refusal. He admitted it was not "a legitimate work for prisoners of war" but wanted to do it anyway as an act of retaliation. He knew that McClellan had reportedly used Confederate prisoners to clear the mines at Yorktown and had also heard that Sherman had used prisoners to dig up torpedoes at Fort McAllister. There is no indication of a response to this by the authorities in Richmond and no indication that the Confederates ever used Federal prisoners in mine clearance.[26]

Given Sherman's hatred of torpedoes, it is interesting that he authorized their use in the railroad cuts along the Atlanta and West Point Railroad. He had become almost obsessed with long-term disruption of Rebel supply lines by that stage of the war. Additionally, his assumption that Black laborers would be used to clear the cuts probably eroded any hesitation he might have had in planting the torpedoes. It apparently never occurred to him that the Confederates would duplicate Union policy by employing prisoners of war for that hazardous work.

The Confederates, for their part, invested quite a bit of time and effort to attack Union rail communications. Giving up territory on a regular basis, especially in the West, they were tempted by the lengthening supply lines of the Federal army. These attacks began early in the war and intensified as it progressed, but it cannot be said that they significantly influenced the course of Union military operations, except in one or two instances. The Federals invested a great deal of resources to protect their lines of rail communications to compensate for the increased level of attacks on them.

The first recorded instance of Confederate mining of rail lines occurred in May 1862, when Maj. Gen. Irvin McDowell occupied Fredericksburg following Confederate evacuation of the town. Haupt reported that "a number of torpedoes with percussion fuses had been placed under the tracks about the depot grounds to blow up trains." Friendly Blacks alerted the Federals to this danger, and Haupt used soldiers to remove them. Unsure that all had been found, Haupt placed a car "heavily loaded with scrap iron" in front of the first train to enter the town, intending to blow up any devices that had been overlooked. It touched off nothing, proving that the clearance effort had been complete. The Federals stored the devices in a brick building detached from the station that had been used to store gunpowder by the managers of the Richmond, Fredericksburg, and Potomac Railroad. One day the building blew up, and all assumed the sentinel had gotten curious but careless in looking at the accumulated torpedoes. "Nothing was ever seen of the sentinel except a piece of his gun at a considerable distance from the spot," Haupt wrote.[27]

We do not know who planted these mines at Fredericksburg, but we do know that before long the Confederates began to expand their attacks on railroads. The group variously known as the Singer Secret Service Corps, Singer Submarine Corps, or Singer's Torpedo Company developed a mine designed for use against locomotives. Singer operatives adapted the detonation device of their submarine mine from a four-pronged rod to a single rod about a foot long. After the charge was planted under a track, the rod would stick up and a trip wire tied to it with the other end of the wire was stretched across the track to a stake on the other side. A passing locomotive was bound to break the wire and set off the charge. This was a simpler system than the one Charles Smith developed, but it also was more obvious.[28]

The Singer group submitted plans for this torpedo to Richmond before August 1, 1863, and it soon was approved. Then the group dispatched operatives to Tennessee in the fall of that year. During October and November, they reportedly blew up eight trains with the torpedo.[29]

It is true that several landmines were used against Union trains in Tennessee in 1863, but it was not always clear as to who did it. Capt. Zere McDaniel, the man responsible for sinking the *USS Cairo* the previous December, is credited with destroying two trains soon after the Army of Tennessee retreated during the Tullahoma Campaign that summer. But newspaper reports credited guerrillas with those attacks and related that the torpedo was set off by a percussion or friction fuse set off by the pressure of the locomotive on the rail. That does not sound like the Singer device and the timing, before July 7, 1863, also predated the deployment of Singer torpedoes.[30]

On October 22, 1863, after the Singer teams had been sent into the field, Brig. Gen. Philip D. Roddey expressed "full confidence in the detachments sent out doing great damage to the road." Undoubtedly, one of these teams was responsible for wrecking an engine on the Nashville and Chattanooga Railroad on the night of October 26. The explosion threw the wreck across the track, completely blocking it.[31]

Soon after the battle of Chattanooga in late November, Brig. Gen. Nathan Bedford Forrest mounted an effort to wreck railroads behind Union lines. According to Lamar Fontaine, Forrest assigned him to command the effort. Fontaine selected twenty-one men, mostly from the Fourth Alabama Cavalry, and used mules to convey "nine large cap-torpedoes to place under the rails, to be exploded by pressure or from a small electric battery, which I had given a thorough test before I started by firing several blank cartridges, at a distance of half a mile from the battery." Fontaine used a telegraph operator to work the electrical detonation system.[32]

Fontaine claimed after the war that his party managed to infiltrate to Rutherford County, where he placed a torpedo inside a tunnel, blowing up a train carrying Federal troops going home on veteran furlough. The tunnel, he argued, collapsed on the train. Then his party escaped but was intercepted on December 14 just south of the Tennessee state line by the Ninth Illinois Mounted Infantry. Fontaine was wounded and captured. The problem with Fontaine's story is that there were no tunnels along the railroad in Rutherford County and no reports of any damage to tunnels along other stretches of the line at this time. Even more damaging to his story was Brig. Gen. Grenville M. Dodge's filing a report of Fontaine's capture, which made no mention of torpedoes. Dodge indicated that Fontaine was a major and a member of Roddey's staff. His party's job was to reconnoiter both the Nashville and Chattanooga Railroad and the line that connected Nashville with Columbia for the Confederates.[33]

We can rely, however, on documents produced by Robert E. Lee that torpedoes played a role in efforts to interrupt Union rail support in Virginia. Maj. John S. Mosby, commander of a partisan unit, conducted an expedition toward Fairfax Court House in early August that encouraged Lee, but the army commander was disappointed that Mosby took so few men along and seemed more interested in attacking wagon trains than the railroad. "If that should be injured," Lee wrote of Maj. Gen. George G. Meade's rail line, "it would cause him to detach largely for his security, and thus weaken his main army."[34]

With Lee's support, Mosby was approached by Alexander R. Boteler, a volunteer aide of Maj. Gen. James E. B. Stuart, about using Rains's torpedoes on the Orange and Alexandria Railroad. Mosby agreed to try it and Boteler requested a number of mines by writing to Confederate Secretary of War James A. Seddon. Boteler assured the secretary that the line was used exclusively for military purposes and relayed Stuart's suggestion that a trained operative be assigned to work with Mosby because the torpedoes were "dangerous things in unskillful hands."[35]

Seddon passed the request on to Col. Josiah Gorgas, chief of ordnance, for his opinion. Gorgas was not enthusiastic. "Unless these torpedoes can be continually replaced as exploded, I doubt the policy of using them at all. To use them once only is to irritate, not intimidate. If, therefore, we have no sufficient command of the vicinity to continue the use of these impediments, it seems to me impolitic to begin their use." But Gorgas assured Seddon that he could supply any number of the devices if called on and would send a man to plant them. John B. Jones, the clerk who had access to this exchange of views, was irritated at Gorgas for stopping the plan.[36]

There was no such hesitancy about the use of torpedoes against railroads in the West, but not every device exploded. The Federals found a torpedo on the Nashville and Chattanooga Railroad in mid-December and dug it up without exploding the thing. Another mine went off without damaging any rolling stock, merely wrecking a short section of track, in mid-January 1864. This alerted the Federals to the need to improve their system of patrolling the line. Union commanders also relied on informants to alert them to the presence of torpedoes. Brig. Gen. Peter J. Osterhaus received word from a man named Harris, who he admitted was "a rather mischievous character," about an attempt to smuggle torpedoes behind Union lines in late February 1864. He took the information seriously enough to send an officer with the news to higher headquarters.[37]

The Confederates continued to press their torpedo offensive in the West up to the moment that Sherman began his campaign for Atlanta. Six weeks before the Federals started their move, Confederate cavalry commander Joseph Wheeler was anxious to equip his troopers with everything they needed for the coming campaign. That included torpedoes because he already was thinking of planting devices on Sherman's railroads.[38]

To most Confederate commentators, the Western torpedo offensive of late 1863 and early 1864 had been a success. The editor of the *Memphis Daily Appeal*, which by this time was lodged in Atlanta, argued that the destruction of trains had been "accomplished at a less cost in men and money than would be required to build, man and equip the smallest gunboat in the Confederate navy."[39]

The Confederates continued their torpedo campaign against Union railroads after the start of Sherman's Atlanta campaign in early May 1864. Wheeler sent small parties of cavalrymen to slip by the Federals, armed with "torpedoes, and other contrivances" to derail a few trains. In one such incident at Calhoun, a group of Confederates drove away a Union patrol and put a torpedo on the Western and Atlantic Railroad, which derailed an empty train steaming north on June 10. They then burned six cars before leaving the area.[40]

Torpedoes were only one weapon employed against Union railroads during the drive toward Atlanta, but they drew an unusual amount of interest. Sherman created the District of the Etowah and placed Maj. Gen. James B. Steedman in command on June 15, charging him with the protection of his railroads from Bridgeport, Alabama, to Allatoona, Georgia. Sherman also contemplated the enemy's use of landmines and offered Steedman his considered opinion on the subject. He felt the use of torpedoes was justified in delaying the advance of an enemy army, but once territory had been occupied, it was a different matter. Using landmines to destroy and kill behind enemy lines "is simply malicious. It cannot alter the great problem, but simply makes trouble." Sherman recommended that Steedman should "test" these devices when found by loading up a wagon or a railroad car with prisoners or with citizens implicated in their deployment and run the vehicle over the mine. "Of course an enemy cannot complain of his own traps," Sherman concluded.[41]

This judgment was vintage Sherman, full of vinegar rather than legal consideration. Steedman did not implement his commander's suggestion. Instead, he declared in general orders issued on June 28 that the placing of torpedoes "to blow up trains containing sick and wounded soldiers and citizens" was barbaric. As of July 7, he ordered subordinates to arrest all civilians within three miles of the railroad line unless they were employed by the government or remained within the picket lines of the various posts along the route. Steedman thereby sought to separate the civilian population from guerrillas, creating, in essence, a "free fire" zone along the endangered rail line. This, with increased patrolling along the track, constituted his policy of protection.[42]

But Steedman's policy failed to rid the line of problems. Capt. Thomas J. Wright of the Eighth Kentucky, US, took charge of forty men from his regiment and the Sixty-eighth Indiana to guard 370 Federal deserters and bounty jumpers onboard a train headed north. The guard rode on top of cars containing ammunition. Wright knew he had to face the possibility of guerrillas firing at the train from the brush as well as a torpedo planted on the track. He heard that a previous train had been blown up near Resaca by a landmine, reportedly "the work of disloyal citizens, or emissaries harbored by them." For Wright, all these threats settled uneasily on his mind. "The thought of riding over tons of powder and striking one of these explosive and inflammable magazines was yet more unpleasant than the fear of rebel bullets."[43]

In July men of the Fourth West Virginia who had chosen not to reenlist boarded a train at Marietta to go home. Halfway between Dallas and Calhoun, the engine hit a torpedo and derailed, wounding some men. The incident caused a delay "of a few hours" for the returning veterans. The Federals often had difficulty knowing exactly who placed charges on the line, and, as we have seen, tended to blame civilians. But a captured Confederate told them early in July that the work was done by detached cavalrymen and scouts sent from the Army of Tennessee.[44]

As the example of the Fourth West Virginia veterans indicates, the odd torpedo attack merely irritated the Federals rather than significantly impeded their logistical support. If Wheeler had sent a large force with hundreds of landmines, the situation might have been different, but small effort in this regard was insufficient to have much effect.

Another major form of military transportation employed by the Federals consisted of hundreds of river steamers, especially on the Western waters. The Confederates began to use torpedoes against that target soon after the fall of Vicksburg on July 4, 1863. Stung by the fall of the Gibraltar of the Confederacy, Seddon mounted a major effort to employ the weapons of the weak against Union shipping on the Mississippi River. Lt. Col. Alfred L. Rives, chief of the Engineer Bureau, sent a list of men authorized in "the special service of destroying the enemy's property by torpedoes and similar inventions" to Edmund Kirby Smith in the Trans-Mississippi and Joseph E. Johnston in Mississippi. The War Department allowed the men to receive cash amounting to half the property value of all they destroyed "by their new inventions" and to retain all guns and munitions they captured "by the use of torpedoes or of similar devices." The Confederates officially mounted a torpedo offensive against river shipping when this document was written on August 20, 1863. The authorized agents included Singer, Fretwell, and Bradbury, members of the same group that had developed underwater and landmines for Confederate service.[45]

As Jefferson Davis recognized, underwater mines seemed ideal to interrupt river navigation, but they were never used by the Confederates on the Mississippi for that purpose. While the activities of the men that Seddon authorized to conduct sabotage is unclear, they apparently concentrated on setting fire to steamers by sneaking on board, disguised as workers. A total of twenty-nine vessels were destroyed in this fashion during the latter half of the war.[46]

The primary torpedo devices utilized in attacking river steamers were those disguised as wood or coal and planted in the fuel supply used by the vessels. The earliest known instance of this occurred not on the Mississippi but on the Tennessee River before the onset of the Atlanta campaign. It was a corollary to the Confederate torpedo offensive against Union railroads of that period.

A handful of small steamers helped supply the Federal concentration at and near Chattanooga during the winter of 1863–1864 by hauling supplies from Bridgeport, Alabama, up and down the Tennessee River. In a conference designed to work out a plan to disrupt that traffic, Maj. Gen. Thomas C. Hindman offered the services of an unnamed officer in his division. That officer approached Capt. I. N. Shannon, the leader of a group of scouts attached to Wheeler's cavalry. Shannon, in a postwar article, explained that he was unable to accompany the expedition but helped in its preparation. He aided Hindman's officer to obtain "fine rifle powder," and recruited five artillerymen and five troopers from Wheeler's cavalry. Hindman's officer led this group of ten men from Dalton carrying the powder, two-inch augers, and two-inch tin tubes that were forty-two inches long.[47]

The men managed to sneak to the Tennessee River ports used by Union steamboats, where they obtained pieces of firewood and bored holes through them. Then they filled the tin tubes with powder and inserted them into the hole, using a pin to keep them there and packing some mud over the hole. Then they slipped these pieces of firewood into the fuel supply used by the steamers and left the area.[48]

Sometime later, the group heard reports that one steamer was sunk and another damaged when the front of its firebox was blown out. The report indicated that the Federals discovered the cause when they found a suspicious piece of firewood and split it open to reveal the charge inside. They blamed civilians who lived near the wood yards used by the vessels. The news was publicized in newspapers, which the group obtained before leaving the area. Shannon later claimed he saw the newspaper report.[49]

The *Chattanooga Gazette* of March 22, 1864, did contain notice of this incident but only confirmed that a steamer had been damaged. The furnace of the *Missionary* blew apart, which led the authorities to examine the wood pile at Sale Creek, upstream from Chattanooga, where they found one of these wood torpedoes. They believed someone who had taken Lincoln's amnesty oath in order to get behind Union lines had been the culprit.[50]

The wood torpedo appeared only once more but, this time, in the form of an alert issued by David Dixon Porter to his vessels operating on the Mississippi River. On May 20, 1864, he warned the Mississippi Squadron to be aware of the weapon, offering a detailed description of its construction that duplicated almost exactly the method Shannon described. The only difference was that Porter's intelligence indicated the Confederates were using sections of gas pipe or shotgun and musket barrels rather than tin tubes. He thought the most likely wood yards where they would be placed in the fuel supply were at Island No. 63 and Island No. 76, but there is no indication that any wood torpedoes actually were deployed on the Mississippi.[51]

Porter also warned his subordinates that the enemy planned to use torpedoes on the Yazoo River. His information indicated they would cut into large trees near the bank of this narrow, winding stream, leaving enough of the trunk intact to keep it standing. Then they would insert explosive devices to blow apart the remaining trunk and topple the tree onto passing vessels. This elaborate and unreliable scheme also was never implemented.[52]

Meanwhile, Confederate Capt. Thomas E. Courtenay developed a different torpedo, designed to be smuggled into the fuel supply of Union vessels. Consisting of a cast-iron casing shaped to resemble a lump of coal, it contained half a pound of gunpowder. Applying a coating of tar and coal dust completed the subterfuge, and sneaking it into a coal barge completed the act of sabotage.[53]

Like Rains, Courtenay promoted his torpedo device where it counted, with Jefferson Davis. By early January 1864, he was in Richmond, lobbying the government for official sanction. "The *coal* is so perfect that the most critical eye could not detect it." Davis liked the weapon, and soon it was being manufactured by the hundreds and issued to a number of Confederate operatives whose missions took them behind Union lines. Soon the Federals became aware of the device through reports by informants. Porter described it accurately in General Orders No. 184, issued on March 20, 1864, to warn his subordinates in the Mississippi Squadron to be wary of what lurked in their coal supply.[54]

Although Davis quietly promoted the coal torpedo and Rains landmines, he was careful not to be seen publicly as an advocate of their use. Surg. Carlisle Terry, who had served in the Army of Tennessee at Stones River and Chickamauga, recalled a story told him by Col. Hypolite Oladowski, chief ordnance officer of the Army of Tennessee. Terry met him at the ordnance depot at Columbus, Georgia, where Oladowski showed him an example of the coal torpedo. Terry accurately described it in an article published in 1890. In fact, Oladowski showed him several sizes and shapes, indicating that the depot at Columbus was manufacturing the device.[55]

Oladowski assured Terry that he had taken some specimens to Richmond and showed them to Davis, but he was stunned at the Confederate president's response. He "was horrified, and furiously declared himself insulted," wrote Terry, that anyone would think "he would be a party to any such unjustifiable mode of warfare." Terry quoted Oladowski's description of the chief executive. His "eye fairly blazed while he gave me such a blessing that I would have been glad to crawl into a rat-hole to get away from him. When he had exhausted his fury he said abruptly, 'Return to your station, sir, this very day.' I firmly believe he would have put me in arrest and preferred charges, but that he did not want the matter to become public."[56]

This story reveals much about Davis and the use of torpedoes. He had always wanted to maintain as much secrecy as possible to maximize the surprise effect of their deployment. But the story also indicates that he fully understood the moral questions surrounding their use as well. "He did not want the matter to become public" was completely right, both to achieve a higher level of effectiveness for the weapon and to deflect negative reaction to the fact that his government officially sanctioned the use of infernal machines.

There is only one confirmed instance of a coal torpedo at work. On November 27, 1864, David Dixon Porter consulted with Benjamin Butler about the proposed Fort Fisher expedition as the two left Bermuda Hundred for Fortress Monroe. They traveled on the *Greyhound*, an army transport that Butler used as his headquarters boat. Soon after starting, Porter noticed "several rough-looking fellows on deck" and told Butler about them. The general ordered the boat docked so the men could be ejected, but they apparently managed to plant a coal torpedo in the fuel supply before they left. A half hour after restarting, the *Greyhound*'s furnace blew up, setting the boat on fire. Porter and Butler barely got off the vessel onto a steamer called the *Pioneer*, later transferring to a tug named *Columbus* on which they made it to Fortress Monroe. Other crew and passengers of the *Greyhound* managed to get onto the *Webster* and were saved, but the vessel was a complete wreck.[57]

Porter had earlier been close to another act of sabotage that he assumed was caused by a coal torpedo. A wharf boat tied up at the Mound City, Illinois, Navy Yard blew up while his flagship, USS *Black Hawk*, was tied to it. The wharf boat was six hundred feet long and sixty feet wide and burned up as a result of the explosion, which took place on the evening of June 1, 1864. The crew of *Black Hawk* moved the vessel away before it was damaged. It makes little sense that a coal torpedo was used, because the wharf boat had no engine. In fact, initial reports indicated that spontaneous combustion of material stored in the paint and oil room was the cause. But Lt. Commander John S. Barnes later reported that the wharf boat had been destroyed by a horological torpedo. This device contained fifty pounds of gunpowder in a box with a detonation system guided by a clock so it could be planted and explode later.[58]

As Barnes asserted in a book published in 1869, many explosions on board Union transports during the war were "unaccountable," but officials suspected the coal torpedo at work. We will never know for certain, because of the secrecy surrounding clandestine operations. That secrecy offered some individuals the opportunity to lay claim to successes that can never be proven.

For example, a man named Robert Lowden claimed after the war that he used a coal torpedo to blow up the transport *Sultana* at the end of the conflict. This was the most famous boat wreck of the Civil War because the *Sultana* was grossly overloaded with released Union prisoners of war when it blew up just north of Memphis on April 27, 1865. Up to 1,500 Federal soldiers lost their lives in the most costly maritime disaster in American history.[59]

Lowden had a checkered past. Using aliases such as Charlie Deal and Charlie Dale, he worked as a painter for three years after the war with a man who later became a newspaper reporter. That reporter told Lowden's story in an article published in 1888. Lowden, according to his telling, had been a Confederate mail carrier and had been arrested several times by Federal authorities but had obtained a pardon from President Andrew Johnson after the war. He had a drinking habit and often told stories to the future reporter during these bouts. One of those stories concerned his activities as a boat burner on the Mississippi River. Lowden claimed to have fired half a dozen transports before it became "too—ticklish a job to set the boat afire and get away from her," as he put it. Because of that difficulty, Lowden's last act of the war was to place a coal torpedo into a pile of coal near the boiler while the *Sultana* was lying over at Memphis.[60]

Lowden's story may sound plausible, but the source of information has to be taken with a large grain of salt. Contemporaries and historians alike have attributed the *Sultana*'s demise to a grossly overloaded deck that contributed to boiler problems. They have discounted Lowden's story as unlikely.[61]

Federal authorities paid no attention to the wood torpedo but were keenly aware of the coal variety. In early May 1865, Assistant Secretary of the Navy Gustavus V. Fox asked Henry Halleck if the Army could obtain specimens of coal torpedoes at Richmond. He was convinced the devices were manufactured at the Confederate capital. John S. Barnes also paid a good deal of attention to the coal torpedo in his book on submarine warfare, published soon after the conflict. The *Greyhound*, carrying two highly placed army and navy officers, garnered a great deal more attention than the little *Missionary* on the Tennessee River.[62]

The horological device was the most sophisticated torpedo developed during the Civil War. Exactly who created it is unclear, but the torpedo was available by the summer of 1864. Capt. Zere McDaniel ordered an operative named John Maxwell, an Irishman hailing from St. Louis and a "bold operator" in Rains's estimation, to use it against Union shipping on the James and Appomattox Rivers. Maxwell worked with a man named R. K. Dillard, who knew the area well. The two infiltrated into Isle of Wight County by August 2, heard of the immense activity at Grant's base at City Point, and decided to go there. They sneaked past Union pickets during the night and arrived before dawn of August 9.[63]

Maxwell left Dillard half a mile from the depot and sneaked in with a box containing twelve pounds of gunpowder and the clock detonation system. He found out that an officer in charge of a barge had gone to shore, so he took advantage of his absence. "I put the machine in motion," Maxwell later reported, and then approached the barge. He gave the box to one of the crew members, telling him the officer had sent it. Maxwell then hurried off to rejoin Dillard and watch as the bomb exploded an hour later.[64]

The saboteur could not have chosen a better barge to blow up. It was named the *J. E. Kendrick* and contained 20,000 to 30,000 rounds of artillery ammunition and 75,000 to 100,000 rounds of small arms ammunition. When the barge blew up, the ammunition also exploded, destroying another barge and a wharf building and their contents. "The scene was terrific," asserted Maxwell from his observation point half a mile away. "And the effect deafened my companion to an extent from which he has not recovered. My own person was severely shocked," he wrote. But both Maxwell and Dillard escaped soon after the blast. Federal Dr. R. B. Prescott witnessed the sudden eruption, which he described as an "immense cone-shaped mass of flame and smoke rising seemingly hundreds of feet into the air, and filled with timbers, saddles, military stores of all kinds; and bodies of men and horses. It was a sight never to be forgotten."[65]

The effect of this blast was frightening. "Five minutes ago," wrote Grant to Halleck at 11:45 a.m., "an ordnance boat exploded, carrying lumber, grape, canister, and all kinds of shot over this point. Every part of the yard used as my Headquarters is filled with splinters and fragments of shell." Several members of his staff were injured, including Col. Orville E. Babcock, who was "slightly wounded in hand." An orderly was killed and two or three others wounded, with "several horses killed." Closer to the source of the blast, witnesses testified to the gory details. "The ded bodys laid arond torn to pices just like pieces of old wood," reported Lewis Crawford of the Third Pennsylvania Heavy Artillery. "Hands and feet and scalps of colored men were rained about the town," wrote a correspondent who barely survived the explosion. "I saw several, and heard of more." This correspondent saw "an object like the entrails of a beef rolled over in the dust. It was recognized as human by a hand and foot attached."[66]

Estimates as to the loss of life vary. Maxwell heard the Federals lost 58 who were killed and 126 who were wounded, for a total of 184 men, while Grant placed the loss at 166 men soon after the explosion. A detailed list of victims published in the *New York Daily Tribune* and undoubtedly contributed by the anonymous correspondent who had barely survived the blast indicated a total of 136 men. Fifty-nine of those listed in the *Tribune*, or 43.3 percent of the total, were Black laborers. Counting white civilians with the Black laborers, the total number of civilians killed and wounded amounted to 73, or 53.6 percent of the total casualties. The rest were military personnel. Ordnance Capt. Morris Schaff was thankful the explosion did not take place an hour earlier, because at that time, the wharf was crowded with people waiting for a boat to Baltimore and the casualty list would have been far longer.[67]

The destruction of material was immense. Schaff's first view of the scene staggered him. "From the top of the bluff there lay before me . . . a mass of overthrown buildings, their timbers tangled into almost impenetrable heaps. In the water were wrecked and sunken barges." The building that had been destroyed by the blast was six hundred feet long. A canal boat positioned between the ammunition barge and the wharf had been filled with worn-out cavalry saddles turned in by Maj. Gen. Philip Sheridan's command. "The explosion sent those old cavalry saddles flying in every direction like so many big-winged bats." One of these flying saddles hit a lemonade vendor in the stomach and killed him. The Federals found a musket nearly half a mile away from the scene of the explosion buried muzzle-first in the ground, standing like a sentinel. While Maxwell estimated the monetary value of the material loss at $4 million, Schaff estimated it at over $2 million.[68]

The blast resulted in some interesting environmental effects. Occurring at the edge of a broad river, it created water spouts, "which, mingling with the powder, cinders and ashes, caused a black, pasty shower, with the other debris," according to the unidentified newspaper correspondent. He noted that the column of smoke could be seen up to forty miles away.[69]

Figure 10.3. City Point Explosion. On August 9, 1864, Confederate operative John Maxwell smuggled a timed explosive device onto an ordnance barge at the massive Union supply base at City Point, which supported Grant's operations against Petersburg and Richmond. It went off with devastating results, killing or injuring up to 184 people and spraying debris over the area near Grant's headquarters. It was the most terrible act of sabotage in the Civil War. This photograph was taken sometime after the explosion, before all the damage to the buildings had been repaired. (NARA, 524967, 111-B-558)

The next day, August 10, up to two thousand men were detailed to clean up the mess caused by Maxwell's torpedo. Flags were raised, and bands played martial music "to soothe the pang of yesterday's desolation," as the correspondent put it.[70]

In the wake of this horrible tragedy, there was no shortage of theories about the cause. Most speculation centered on careless handling of ammunition boxes on the *J. E. Kendrick*, while others thought it was a submarine torpedo. A few correctly assumed a spy or saboteur had been at work, while others picked a long shot by arguing a Confederate artillery piece on the north side of the wide Appomattox River had fired a projectile onto the barge.[71]

Maxwell continued his attacks on Union shipping after the City Point incident. He leagued with a party of operatives, apparently from the Confederate navy, and helped to capture the transport *Jane Duffield* in the Warwick River on September 17, 1864. Sometime later, the Federals were able to identify him, and by December 1864, he had returned to Richmond. The case against Maxwell was sealed when Dr. R. B. Prescott found his December 16 report of the City Point sabotage in June 1865 and sent it to the authorities in Washington. Halleck ordered the arrest of Maxwell, Dillard, Rains, and McDaniel "if they [could] be found" and forwarded the report to Secretary of War Edwin M. Stanton. There is no indication that Rains, the most prominent man on this list, was ever arrested and equally no indication that the others were tracked down.[72]

While Dr. Prescott's discovery of the Maxwell report solved the mystery of what caused the City Point explosion, Morris Schaff, in his own way, closed the books on it as well. In charge of the ordnance depot at City Point, he was responsible for every penny of government ordnance property at that massive supply base. On leaving the depot, he had to document the disposition of everything and, inevitably, there were items not accounted for by proper paperwork. Schaff just listed those items as blown up on August 9 "and let the explosion balance the books."[73]

Despite all the ingenuity and effort expended in using torpedoes to target communications during the Civil War, some men thought in terms of using them to enhance transportation rather than destroy it. The Federals devised a submarine mine for the removal of obstructions in river and harbor channels. A horological device, set up to be detonated by a clock, was to be washed toward the obstruction by the tide and explode when near it. The method sounds unreliable, and there is no evidence that it ever was used. Although it would be possible to theoretically coordinate the timing of the device with the known movement of the tide, there always would be variation in every tidal movement that could derail carefully laid plans. But at least someone was thinking of using torpedoes for constructive purposes.[74]

As with their use of landmines in main-force operations in the field, the Confederate use of torpedoes in targeting communications failed to have a significant effect because they used minimal resources in the effort. Far more locomotives were derailed by loosening a rail than by planting a torpedo, and the former tactic was far easier to do than obtaining an explosive device and learning how to place it. Moreover, for every offensive tactic, there was a defensive measure. The Federals quickly learned about the coal torpedo and took measures to lessen their exposure to it. The City Point explosion was devastating, and if it had been duplicated many times, the accumulated effect would have been widely felt. But in a sense, it was a fluke. Maxwell happened to arrive at exactly the right moment to plant his device on the best barge for his work. He did not even try to duplicate this tactic but instead refocused his efforts on capturing vessels, which produced little in the way of benefit to his cause. The Confederates had the technology and a handful of bold operators, but they did not have enough material, men, or vision to launch a large-scale torpedo campaign against Federal shipping.

Chapter 11

Developments in the United States after the Civil War

"The Weapon the Father Prepared, May Turn Against the Son"

The use of torpedoes in the Civil War created a great deal of resonance in the postwar era. Technically, the latter war months combined with the immediate postwar period to witness a good deal of work toward developing a reliable electrical system of detonating landmines. Moreover, anger over their employment continued to be strongly felt by Northerners after the war ended. Discussion of the tactical effectiveness continued, and Gabriel Rains engaged in a lifelong campaign to establish himself as the great man of torpedo warfare, even offering his services to the US government. Victims of torpedoes also had a lifelong job—dealing with the physical and emotional effects of their sudden exposure to modern warfare.

The Civil War occurred at a time when improvement in the detonation of explosives was progressing in a marked way. The old fougasse had been set off primarily by a lanyard with an observer at one end of the line ready to tug on the connection that set off the device. Immanuel Nobel developed an effective system of victim-activated landmines for the Russian defense of Sebastopol as had Gabriel Rains in Florida a few years before. Rains continued his self-activated mine work in the Civil War, but some torpedo men preferred electrical detonation. Such systems had been deployed at Columbus and Fort Fisher; they required an operator to keep watch for the right time to apply electrical current to the mines.

Both galvanic (also termed chemical) devices and electromagnetic devices were used to set off gunpowder through the middle of the nineteenth century, but the latter held more advantages than its competitor. By the time of the Civil War, improved electromagnetic devices were capable of use for military purposes. While chemical batteries could set off torpedoes, they provided a constant flow of electricity, most of which was wasted as the operator waited for an opportunity to set off the device. Electromagnetic systems provided a brief surge of electricity when the operator turned a crank that rotated a pair of coils around a stationary magnetic armature to send an electric current through insulated wires to the fuse.[1]

L. C. Baker, a telegraph operator working with Confederate Maj. Gen. Thomas C. Hindman in Arkansas during the summer of 1862, demonstrated the limitations of galvanic batteries for torpedo work. He filed a detailed report on the Grove battery, a chemical device widely used in telegraphy. "I could not bring sufficient battery to bear on the fine wires to deflagrate them," he wrote. "It would require from fifty to seventy-five cups of Grove battery to explode a mine at a distance of 1,000 yards or a mile. I could not place more than twelve or fifteen in circuit. There are not exceeding thirty or forty cups, that I know of all told, west of the Mississippi." Insulated wire also was severely limited. Baker was so busy using his battery for telegraphic work that he barely had time to experiment enough to know that it would be unwise to attempt a torpedo explosion with less than fifty to seventy-five cups.[2]

The electromagnetic device did not suffer from the limitations of the chemical battery when it came to exploding torpedoes. But Gabriel Rains had no faith in any electrical methods of setting off mines. "Torpedoes to be fired by electricity are not reliable and are not to be compared with those self acting," he wrote in the "Torpedo Book." Rains never explained his reasoning, but he had a self-motivated interest in downplaying electricity in favor of his favorite method of victim-activated mines.[3]

George W. Beardslee of Brooklyn thoroughly disagreed with Rains. He developed a twist on the electromagnetic device by keeping the coils stationary and revolving the magnets to produce a machine designed for use in electroplating by 1859. Beardslee increased the number of stationary coils to twelve and added a switchboard so the operator could "key in any number of coils to make up the generator's output." Despite its capabilities, Beardslee's machine was not widely adopted for electroplating. Firms involved in that business tended to prefer chemical batteries.[4]

Frustrated, Beardslee turned to adjusting his device for use in the telegraph industry and achieved some degree of success. He filed for a patent on his invention designed for powering telegraph lines on August 4, 1863. A few days later, he filed for another patent for his device designed for exploding gunpowder, on August 18. Both applications were relevant for the war being waged by the North. The latter device supposedly could set off up to twenty-five charges at once. Beardslee also developed his own fuse for it and crafted a cast-iron torpedo as well.[5]

Beardslee's machine was adopted by the US Signal Corps to power its field telegraph system. But it did not satisfy the army's needs, providing current for no more than five miles of wires rather than the promised twenty-five miles. Field telegraphers preferred the steady current of galvanic batteries and soon replaced the Beardslee machine.[6]

Frustrated again, Beardslee worked to win army approval for his method of exploding land and water mines. He convinced Chief Engineer Richard Delafield to approve an official trial. On January 17, 1865, Delafield appointed a board to examine the device and witness trials at the US Military Academy at West Point. Using three machines of differing sizes, Beardslee set off several charges on January 24. He returned to West Point on February 2 to arrange twenty of his cartridges along a wire eight feet apart and set seventeen of them off simultaneously, deliberately not firing the last three to demonstrate he had control over the process. The board wanted more trials, so Beardslee returned on April 17, a week after Appomattox. In addition to exploding underwater charges, he placed sixty eight-inch shells on the ground in sets consisting of twenty each, every shell eight feet apart. He was able to explode only the first set of twenty landmines, but none of the shells in the other two sets went off. Upon investigating, he found the insulated copper wire arrangement was faulty.[7]

Nevertheless, the board was favorably impressed. Even after the January demonstrations, it had filed a tentatively optimistic report, favoring electrical detonation because it dispensed with acids and liquids. Beardslee's device was "very portable, without risk of accident." His system was "certainly very promising and [could] not fail to be of great practical utility in the firing of mines and torpedoes at considerable distances and depths." The board did not go so far as to urge the adoption of the system, merely reporting on the technical value of it for army operations.[8]

Beardslee had a close rival in Taliaferro P. Shaffner, a self-taught inventor and telegraph promoter who had been involved in the creation of the North American Telegraph Company. Shaffner spent several years in Europe working on what he termed "the application of Torpedoes and Military Mines for Offensive and Defensive Measures in war." Most of his time had been spent in Russia, Denmark, and Sweden, and there is some indication that he was active for one side or the other in the Danish War of 1864. Shaffner claimed that a total of nearly $250,000 had been spent in his European endeavors, although he did not provide details. A fervent advocate of the Union cause despite his Virginia birth, Shaffner returned to the United States. "I abandoned my great enterprises in Europe and purchased about ten thousand dollars ($10,000) worth of materials suitable for Torpedo and Military Mining purposes." He arrived early in 1865.[9]

Even before his material cleared customs, Shaffner lobbied Ulysses S. Grant in late February 1865, about the same time as the Beardslee trials at West Point. He spent a couple of days briefing engineer officers about his system. On February 27, Shaffner assured Grant that he could set off at least six torpedoes at the same time at almost any distance. He especially praised his fuses, which were "exceedingly sensitive and will explode without the possibility of a failure to ignite at the precise moment." The fuses were guaranteed to ignite all the powder in each charge to produce the maximum effect. Shaffner required no compensation for his services. In return, Grant endorsed Shaffner's system and arranged with Secretary of War Edwin M. Stanton to bring his material to City Point for demonstrations.[10]

Stanton authorized the release of Shaffner's material from the Custom House in New York City without dues on March 3. Meanwhile, Grant had shown Shaffner the fuses Comstock brought from Fort Fisher, and the telegraph promoter assured him they had been made at Woolwich Arsenal in England. Secretary of State William Henry Seward reported that Frederick Augustus Abel was said to be working at Woolwich, so the assumption was that they were Abel fuses.[11]

Grant appointed a board to examine Shaffner's arrangements, and by March 22, the board filed a positive report. "This system can undoubtedly be made of incalculable value," Grant wrote as he endorsed the board's recommendation. It was an all-purpose system, capable of firing mines underwater or on land, but Grant especially thought it could be useful in harbor defense against a European naval force.[12]

It could also be useful at Petersburg, where the Federals had already exploded a large mine thirty feet below the surface to blow up a salient of the Confederate works on July 30, 1864. Six days after Grant's endorsement of the board report, Shaffner informed the general of what he needed to equip the Federal army at Petersburg for more underground mining projects. Grant authorized the purchase of everything the same day, March 28, 1865.[13]

But five days later, the Petersburg campaign came to an end with Lee's evacuation of his lines and the fall of that city and Richmond. There was no longer a need for Shaffner's system. War's end left the promoter high and dry. A year later, he appealed to President Andrew Johnson for some kind of government appointment, explaining that he had devices useful to national defense, such as one that used electricity to find the position of a ship up to three miles away. The inventor overstated his role in science by claiming that electricity was "a branch of physics developed by me in the art of war, more than, perhaps, all other men." Grant endorsed Shaffner's application for a job but admitted that he could not see how a position might be created for him, and the appointment never came to be.[14]

The Civil War boosted efforts to improve the efficiency of electrical detonation of land and water torpedoes and underground mines alike, but none of those efforts resulted in actual use of improved chemical or electromagnetic devices. Beardslee and Shaffner arrived too late in the war for their machines to be used. In fact, most of the landmines deployed during the war were self-activated devices triggered by contact with a person. Although electrically detonated torpedoes were rigged up in two instances on land, not a single one of the devices were exploded at those locations.

Beardslee did not give up. He took his system of electrical detonation to Europe, and the English were impressed. "The late civil war in America has occasioned the practical application and modification of these engines of warfare in a great variety of forms," noted the *Illustrated London News*. Thus Beardslee was received as an expert in his field when he arrived loaded with wires, detonators, and fuses. The editor recognized that Beardslee's electromagnetic device was superior to the Grove galvanic battery for detonating gunpowder. It was portable and easy to use.[15]

Beardslee informed the English of his West Point trials and also told them of other demonstrations he had conducted in the United States. He claimed to have connected his device to the telegraph line between New York City and Washington, DC, in August 1864, setting off a charge of gunpowder first in the capital and then in New York by electrical currents. That was reportedly a distance of 240 miles. Earlier than that, according to Beardslee, Grove's battery had been tried in a similar experiment on the telegraph line between Washington and Baltimore, a distance of only forty miles. The galvanic battery managed to explode the powder but needed "the combined power of no less than 150 pairs" of cups to do so.[16]

The *News* editor described Beardslee's fuse as "a small wooden cylinder" three quarters of an inch in diameter and three inches long. It held combustible material, ignited by the electrical current, in one end. These fuses were employed in demonstrations staged at the mouth of the River Medway on October 11, 1865. Beardslee prepared for a week but concentrated only on exploding underwater charges. The trials were a success.[17]

In all his demonstrations, Beardslee managed to prove the technical utility of his electromagnetic system. Whether he was successful enough to secure a contract with the English government is not known, but his was not the only competition to the Wheatstone device that was then popular in England.

In the post–Civil War era, electrical detonation failed to become important for setting off landmines. But for detonating underground mines in siege warfare, it became vitally important. This was the area in which people like Beardslee and Shaffner could have made their inventions felt. Dispensing with the old powder train, electricity held the potential to make underground military mining as safe as it would make commercial mining. But Federal forces only moderately engaged in military mining during the Civil War, and all projects had ended long before Beardslee and Shaffner had offered their services.

The Civil War did spur a little-known twist in extracting petroleum from wells, and torpedoes played an important role in that process. It was developed by Lt. Col. Edward A. L. Roberts of the Twenty-eighth New Jersey, who had a checkered military career. He was charged with drunkenness while on dress parade in November 1862, and his court martial had not yet rendered its verdict when the regiment was thrown into battle at Fredericksburg a month later on December 13. Col. Moses N. Wisewell was shot, and Roberts took command of the regiment for the remainder of the engagement, performing well. That did not prevent the court from later issuing a guilty verdict, and Roberts was dishonorably dismissed from the army on January 12, 1863.[18]

Roberts soon made the most of his wartime experience. He had noticed during the battle at Fredericksburg that Confederate shells falling into a canal seemed to push the water sideways with great force. This led him to think of using explosives in the shafts of oil wells to shatter the rock layers and release more oil to be pumped out. He drew up plans for what came to be known as the petroleum torpedo in November 1864, making six devices. When he approached owners of oil wells near Titusville, Pennsylvania, the proprietor of the Ladies' Well allowed him to try the experiment. Roberts exploded two torpedoes in his shaft on January 21, 1865, and it worked so well that he formed a company and filed a patent application on April 25 of that year. Roberts used fifteen to twenty pounds of gunpowder in the torpedo, filled the shaft with water, and lowered the device until submerged. The water provided what a modern historian has called "fluid tamping" that would "concentrate the concussion and more efficiently fracture surrounding oil strata." Roberts continued to promote his system for nearly two years until achieving success with the Woodin Well, which had never produced oil. After one torpedo in December 1866, it released twenty barrels a day. A second torpedo in January 1867 increased output to eighty barrels a day, and demand for his system skyrocketed.[19]

Roberts conducted business with a hard attitude, charging exorbitant fees of $100 to $200 per torpedo and insisting on a royalty of one-fifteenths of the increased oil production. This led to many cases of illegal use of his system, and Roberts spent $250,000 to detect violations of his patent and file lawsuits to stop them. That patent expired in 1879, giving his competitors free rein to use his system. Until then, the Roberts Petroleum Torpedo Company had a legal monopoly. He tried to improve the system as he went along. Initially the torpedo was set off with a percussion cap lodged on top of the canister, and a weight was dropped down along a line to detonate it. This, however, led to premature explosions. Then he placed the torpedo inside a larger metal tube with sand filling in the space between. The fuse was on the inner canister and a blasting cap on the outer one. Why he did not experiment with electrical detonation, which probably would have been the safest thing to do, is not explained. Perhaps it was too costly in the mind of this vigorous businessman. Nitroglycerin, which had been developed by 1846, was first used successfully in 1861 and, by the time of Roberts's discovery, was spreading in commercial blasting. Roberts began to manufacture his own nitroglycerin by 1868 for use in his petroleum torpedoes, claiming it was too dangerous to ship it even by railroad.[20]

In a purely technical sense, the petroleum torpedo cleaned the oil well of paraffin wax buildup and opened fissures in the surrounding rock strata. Both tended to increase the seepage of oil into the well and thus increased output. It was the predecessor to the modern practice of hydraulic fracturing, which uses pressurized liquid to accomplish much the same object. Beginning in 1947, it has become so widespread that more than 2.5 million instances of its use had been recorded by 2012, nearly half of them in the United States. While industry argues it is necessary to fracture in order to extract gas and oil, environmental advocates highly criticize it for, among other things, increasing the leakage of methane gas.[21]

Technology often produces moral, humanistic, and environmental problems that tend to trouble society long after its use has been established. Efforts to deal with those collateral problems have to play a sometimes-deadly game of catch-up, often losing the race to curb excesses. One of the most persistent results of Confederate landmine use in the Civil War was the Northern disdain for what was perceived as a barbaric weapon, callously employed by their now subdued enemy. The end of the war did not end Northern anger, but ironically, that anger did not prevent the government from adopting a torpedo program of its own late in the conflict and after the war.

William T. Sherman's abhorrence of landmines has already been documented. He was particularly incensed by the Pooler Station incident and by the landmines at Fort McAllister because he had personally witnessed their effects on his men. Sherman never forgot the incidents and never forgave the Confederate perpetrators of what he considered a crime.

Many years after the war, Sherman visited Denver, Colorado, and was approached by a man who was missing his right leg. Sherman "in his quick nervous way scanned his features for a moment," then offered his hand and told the man how vividly he remembered Pooler Station. The man was Francis W. Tupper, now a deputy clerk in the US court in Denver. Tupper was amazed, he told a reporter, that Sherman recalled so well the entire incident, even though the two had not met since December 9, 1864. They had a heartfelt conversation about that day.[22]

In his memoirs, published in 1875, Sherman made it plain that he still viewed landmines as unjustified weapons. In describing the Pooler Station incident, he wrote of using Confederate prisoners to locate the mines by marching them over the roadbed. Former Confederate Gen. P. G. T. Beauregard was incensed by this reference to forcing prisoners to dig up torpedoes after reading the memoirs. He denied Sherman's assertion that this was done in retaliation for Confederate use of Union prisoners in similar circumstances. Beauregard could not recall any instances where it would have been possible to use Federal prisoners this way, entirely forgetting the fact that he had wanted to do so while reopening the Atlanta and West Point Railroad after the fall of Atlanta. He called the whole idea of using prisoners of war in this way an example of "shocking inhumanity." But Beauregard admitted that after the fall of Savannah, on hearing reports of Federal use of prisoners to clear mines, he had requested permission from Brig. Gen. John H. Winder, Confederate Commissary of Prisoners, to detach some Union captives "to be employed in retaliation should the occasion occur." Winder replied that he could not grant permission, as the War Department refused to allow prisoners to be used that way.[23]

Beauregard did not make public his reaction to Sherman's memoirs, writing all this in a letter addressed to an unknown correspondent in 1878. Ironically, this letter was printed in a newspaper article that found its way into Sherman's papers, so the Union general was fully informed of Beauregard's views.

It was all well and good for Confederates to protest the use of their men to clear landmines. But they never so warmly criticized their own people for planting the mines in the first place. This point was not lost on Northern veterans. When Gilbert Adams Hays compiled material for the history of his regiment, the Sixty-third Pennsylvania, he included a comment on Southern criticism of McClellan's use of prisoners to clear up the mines and booby traps at Yorktown. "They had not a word to say against the savage barbarity of those who planted these deadly engines in the pathway of the army."[24]

Gabriel J. Rains never regretted his career as a torpedo man. In fact, he became—if anything—more vigorous in promoting landmine warfare after Appomattox than before. Rains lived in Atlanta and then Augusta after the war, employed by a fertilizer company in the latter city for many years. In about 1874, he wrote his "Torpedo Book," an effort to document his role in the development of landmines. It probably was similar to, though longer than, the book manuscript he had written during the war and which was so influential in convincing Jefferson Davis to adopt torpedoes as official Confederate policy. That earlier manuscript has never surfaced.[25]

Penned while living at Augusta, Rains poured everything into the "Torpedo Book." He headed one section of the manuscript "My greatest discovery in War," referring to "Sub-terra Shells or Land Torpedoes." Rains grossly exaggerated when he argued that "this is regarded as the 'ne plus ultra' of modern inventions for warlike purposes, for by it we can turn & checkmate mighty armies, and so demoralize soldiers as to paralyze all efforts for successful warfare." His devices would "render invasion impossible." In short, Rains continued the hugely overblown view of the tactical effectiveness of his weapon that he had held during the war. He based this exaggeration, as we have seen, on reports gathered usually from civilians who had, at best, second-hand information. For example, he cited from a letter written by an unnamed lawyer living in Williamsburg that the small number of torpedoes Rains planted on the Williamsburg Road on May 6, 1862, had caused the Army of the Potomac to hesitate three to four days before resuming its march along a different road. He felt no need to verify this absurd claim or seek more reliable evidence.[26]

"Pure philanthropy led to this invention," Rains continued in his typically self-important manner, "for if by it we put a stop to all aggressive forays, hostilities become impossible." He even predicted that mankind would stop fighting wars because of his invention.[27]

Rains entered firmer ground when he argued that war itself was "but murder legalized on a large scale" and that moral arguments against landmines were therefore moot. "Each new invention in the material of war has been assailed and denounced as barbarous and anti-Christian, yet each in its turn notwithstanding has taken its position by the universal consent of nations according to its efficiency in human slaughter." In that sense Rains was writing sensibly about the history of military technology, but he was far too optimistic in predicting that torpedoes would so demoralize troops as to render them useless as soldiers.[28]

The danger posed to civilians by landmines became part of Rains's thinking, but he never truly reconciled it with his vigorous promotion of torpedoes. He wrote vaguely in the book that "ordinarily, the use of these shells for such purposes have been refused—lest thereby noncombatants, as women & children be killed." Yet he went on to describe how to make it difficult, if not impossible, for the enemy to locate and remove the shells, which would only increase the danger to noncombatants after the fighting ended.[29]

Rains was aware of the moral argument against his weapon but never succeeded in efforts to counter that argument. Merely noting that new military weapons in the past had often been criticized is not a reason for accepting the next one. Expressing concern for unintended victims on the one hand but continuing to find ways to secret these devices in ways that inevitably endangered them exemplified the fact that Rains could find no way to avoid hurting people he did not want to harm. And arguing that the most important aspect of his invention was to demoralize rather than kill enemy soldiers fell flat because, as we have seen, widespread and paralyzing demoralization had never occurred among Union soldiers faced with a torpedo threat.

Basing his evaluation of the tactical use of torpedoes on his desperate hope that demoralization alone would turn back advancing armies, Rains advocated in his "Torpedo Book" that one thousand men should be deployed to plant landmines in the path of invading columns. Even the possibility that a road could be mined would make soldiers hesitate to advance, he assumed, and the certainty that it was mined would end any forward movement. While a thousand artillery shells could be fired without effect in a battle, planting those shells as torpedoes made sure that every round counted in the "work of destruction."[30]

Rains worked out the details of transporting and placing landmines in the face of advancing armies. They could be carried in each end of a bag designed to be slung across the back of a mule. The bag could also be carried over a pole by two men. When planting them, it was best to "kneel on the right knee." How far apart to plant was a matter not yet fully worked out. Rains assumed that even if not connected to another shell, one sub-terra would cause neighboring shells to explode too. "How far this influence extends is not yet determined," he wrote, "but [it is] supposed to be about as many feet distant as the number of inches in the diameter of the shells." To ensure multiple explosions, it was necessary to attach a wire to as many mines as the operator wished. If the operator used an electromagnetic device, "an army may be destroyed in a moment of time at the will of one man." Rains suggested that planting the shell deeper than was usual during the war, as deep as two feet below the surface, would render it far more difficult for the enemy to find. Placing a stick with the bottom end over the primer and the upper end just below the surface would enhance detonation. In addition to mining roads, Rains suggested that torpedoes should be planted between the ties of a railroad. He envisioned placing pieces of wood so that one end would be pressed by the engine wheel and the other would press the detonator. Mining beach landings and river crossings also came within Rains's vision of tactical deployment.[31]

Finally, Rains provided details about how to plant an extensive mine field as a semipermanent obstruction. After placing the shells, he advised attaching a foot-long piece of string, one end tied to the shell, and at the other end, a small piece of red flannel should be attached. The flannel would stick up above the surface of the ground, but because it was only as big and long as a finger, it would be inconspicuous. The operator should make a record of the planting to identify exactly where the field was and even blaze nearby trees. Rains advocated this method because he recognized that it might be necessary to remove the shells. To preserve his role in the improvement of torpedo technology, Rains described in detail how to make sensitive fuse primers for electrical detonation.[32]

Rains produced in his "Torpedo Book" the first manual of torpedo warfare in world history. It remained unpublished until historian Herbert M. Schiller edited it in 2011. Rains remained on safer grounds when discussing the techniques of planting and assessing the technology of the sub-terra shells, but his assessments of the effect of these weapons on the enemy were wildly exaggerated.

At about the time he was writing the book, Rains tried to secure an appointment to teach his doctrine at West Point. Writing to President Ulysses S. Grant, Rains referred to "our old friendship" as he urged the chief executive to install training in torpedo warfare at the US Military Academy. He believed every advanced army in the world was already using torpedoes and wanted the United States to catch up. His "Torpedo Book" represented "almost the study of a life time" and explained everything officers and cadets would need to know. He offered to teach the course himself, "giving my experience to the advancement of our service[,] enhancing the peace prospects of mankind in rendering wars measurably impossible by such formidable inventions."[33]

Grant passed this letter on to the secretary of war, former Union general William W. Belknap, who was not impressed. "If the introduction of this study should be contemplated there is no necessity for securing the services of a Rebel General," he wrote. Ironically, Belknap did not even know that such a course of instruction had long before been instituted at West Point. Col. Thomas H. Ruger, superintendent of the academy, assured Belknap that the cadets were instructed to know the general principles, the different types of mines and detonation devices, the method of placement, and the tactical situations in which it was useful to deploy them. "I give special instruction with a view to make entire classes of cadets competent to practically supervise the manufacture of and use of torpedoes," Ruger reported.[34]

The man responsible for teaching cadets about torpedoes was Capt. Oswald H. Ernst, instructor of practical and military engineering at the academy. He authored a manual on military engineering, which was published in 1873, a year before Rains wrote his letter to Grant. Ernst saw landmines as "a valuable obstruction from its effects, both physical and moral, but especially the latter." The two detonation systems included a contact device, which he called the "sensitive torpedo," and electricity. Ernst considered the contact mine "a dangerous weapon to use, as casualties will constantly occur from accidental explosions both during and after the planting." But the disadvantage of electricity was the difficulty of timing the explosion "at the proper moment."[35]

Frustrated in his rather naïve assumption that the US government would welcome a former Confederate general as an instructor at its military academy, Rains took his campaign to the public by writing an article for the *Southern Historical Society Papers* in 1877. Other nations were studying torpedo warfare as theory, but Americans had had the most extensive experience with it during the Civil War, he correctly argued. As in the "Torpedo Book," Rains pursued the line that his invention would lead nations to renounce war in favor of international arbitration of their difficulties. People would protest, surely, but eventually everyone would convert to their use because of the crowning benefits of "universal peace." He recognized that the use of underwater mines had garnered more international attention than that of landmines but was certain the latter would catch up with the former in visibility and widespread use.[36]

After making his case in a lucid, if unconvincing, manner, Rains then spent a good deal of space in his *Southern Historical Society Papers* article describing the Fort King incident. Here his mind began to wander until much of his discussion of the first-ever planting of a booby trap in American history became gibberish. This gives us an indication of why some people even in the Civil War viewed Rains as little more than a crackpot.[37]

In the increasingly obsessive effort to secure his legacy, Rains wrote a long letter to W. T. Walthall, a former Confederate officer who was helping Jefferson Davis gather material for the former Confederate president's history of the war. Rains concentrated on the sensitive primer as the key to his concept of torpedo warfare. Not only did it require the slightest pressure to detonate, it could easily be inserted into and out of the sub-terra shell. "This primer is the secret of the success with Torpedoes both on land and water," making them "the most formidable destructives ever invented, whose efficacy was tested by me and a colonel of my command by experiment." Again, he exaggerated the effect of his weapon by arguing that the torpedoes arrayed before Fort McAllister actually repelled the Federal attack. But then most members of the garrison lost their nerve and retreated into bombproofs, allowing a second attack to succeed. As we have seen, this was completely untrue. "Thus Nations have been furnished with a new weapon of destruction," Rains concluded in his letter to Walthall. The landmine not only made invasion more difficult, but it advanced "another step among Christian Powers to universal peace."[38]

Davis needed little encouragement. He had fully embraced torpedo warfare in 1862, making it official Confederate policy, and never changed his mind. In *The Rise and Fall of the Confederate Government*, published in 1881, he gave Rains full credit for planting the mines at Yorktown, something that Rains himself never admitted. Davis gloated over the "terror" inspired among McClellan's troops when they stumbled upon these devices. He praised Rains for the Williamsburg Road planting, which stopped the Army of the Potomac in its tracks, and credited the sensitive primer as an important invention.[39]

Davis provided nearly all the public admiration Rains would receive as the progenitor of torpedo warfare. Few others offered any praise of his work, and Rains probably did not even see Davis's book. In 1877 he had moved from Augusta to Charleston, where he secured a position as a clerk in the US Army Quartermaster Department. Three years later, Rains moved to Aiken, South Carolina, seeking a drier climate because his Seminole War wound began to trouble him. He died there of heart disease on August 6, 1881, the same year that Davis's book appeared. Thirteen years later, one of his daughters extracted material from a diary and scrapbook she found in her father's papers for an article published in *Confederate Veteran* that described the Fort King incident. The editor entitled the article "The Invention of Torpedoes: Gen. Gabriel J. Rains, of South Carolina Bears the Honor." One of his daughters, Katherine Rains Paddock (who might have supplied the material for the *Confederate Veteran* article), donated his "Torpedo Book" to the Museum of the Confederacy in 1926.[40]

The appearance of Davis's history of the Confederacy brought out latent jealousies within the ex-Rebel torpedo community. Hunter Davidson, who had worked mostly in submarine mines, was irritated that his name was not mentioned. Davis praised "the performance of his dear friend, General Rains, whom he favored in every way possible." But Davis did not refer to "the amusing performances of General Rains in the Florida Seminole War with torpedoes and sensitive fuses concealed under blankets to attract and catch the Indians. But the biter was bit, and the Indians caught him and peppered him with lead. He was daft on sensitive fuses, and his experiments were generally disastrous."[41]

Those who worked with submarine mines began to create their own legacy in publications after the war. Davidson wrote an article for the *Confederate Veteran* about them, and Dabney Herndon Maury testified to their effectiveness in the defense of Mobile. To be fair, Rains had given attention to the importance of submarine mines in his *Southern Historical Society Papers* article but without giving credit to others for them. Estimates as to how many Union vessels were sunk or damaged by underwater mines during the war varied, but the most authoritative count came from torpedo historian Milton F. Perry. He listed a total of forty-three vessels beginning with the *Susquehanna* in the Wright River of Georgia on February 14, 1862, and ending with the *R. B. Hamilton* in Mobile Bay on May 12, 1865.[42]

Although he had worked at times with underwater mines, Rains never became associated with them in the public mind as he had with the use of landmines. As we have seen, even among ex-Confederates, his legacy was checkered. While some Southerners referred to him as the man who "invented the 'torpedo system,'" others did not take him very seriously. Edward Porter

Alexander wrote after the war that "the use of the torpedo was an old hobby with" Rains.[43]

Whether one viewed Rains as a pioneer or a quack depended a great deal on how they viewed torpedo warfare. There is no denying that the general played an important role in the development of that mode of operations. He was the first man in history to improvise a landmine from an artillery shell and primer, creating a booby trap rather than an open element of defense for a fixed asset like a fort. He was not the first Confederate to plant torpedoes or lay booby traps in the Civil War, but he was responsible for the first incident that garnered a great deal of attention from Unionists and Confederates alike. Caught by the publicity of the Yorktown incident, Rains aggressively became a public and vigorous proponent of an expanded concept of torpedo use. He succeeded in this endeavor only because Jefferson Davis was easily convinced and adopted his concepts as official policy. Rains became Davis's torpedo troubleshooter, darting about to threatened areas to plant his device. He also headed the Torpedo Bureau, established at Richmond to supply explosives and technical experts when called on to do so. In an administrative sense, he had accomplished a great deal to promote the acceptance and use of his device.

In a technical sense, Rains also succeeded in developing a system that worked very reliably. He invented nothing, simply using preexisting artillery shells, but his improvement of the basic percussion fuse into a more sensitive primer was significant. That primer was more reliable than an ordinary fuse because it could be set off by only seven pounds of pressure. The system of planting and his new concepts concerning how to arrange the torpedoes and identify their location were all new and quite effective.

But in a moral sense and a tactical sense, Rains either failed or his concepts were far removed from reality. He was highly aware of the controversy surrounding his system and tried to deflect criticism by citing history, using a bit of philosophy, and vainly arguing that his system could curtail future wars through its demoralizing effect on enemy troops. None of those arguments went very far with his critics. Although Rains was aware that innocent people could be hurt by his devices, he never came up with an answer to that problem. His wild predictions about the tactical effectiveness of his devices—that they would render the enemy impotent through terror—were disproved from the very beginning of torpedo warfare. Rains lived within a world created by his desperate hopes, ignoring evidence that contradicted them and eagerly grasping any bit of news that seemed to support them.

Historians Michael P. Kochan and John C. Wideman have stated that Rains "developed the sub-terra shell into a masterpiece of terror and tactics."[44] But that is not so. He developed a landmine that technically worked well, but it fell very far short of terrorizing the Federals and never affected their tactics in any appreciable way. Despite his technical expertise, Rains was, to a significant degree, what he appeared to many of his compatriots: a crackpot.

Rains, however, started a process whereby acceptance of the use of landmines grew in the United States. As we have seen, initial Confederate resistance had quickly evaporated. Joseph E. Johnston had initiated an inquiry into the Yorktown incident after he learned of it through Northern newspapers. There is no evidence that anything came of that inquiry, and Johnston never stated publicly that he disliked the use of landmines—he was irritated that they might have been used without his knowledge and offered no protest that we know of when he discovered they had been employed. More than a year later, as we have seen, he urged Rains to plant sub-terra shells to aid his evacuation of Jackson and called on the general for more torpedo help during the Carolinas campaign.

Ironically, Johnston even forgot that landmines had been used at the time of the operations at Yorktown. Twenty-three years after the event, he read in Davis's book of the plantings on the Williamsburg Road on May 6, 1862, and stated in an article published in *Century Magazine* that he did not believe it was true. This prompted the editors of the magazine to dig into the records and offer more than enough proof from Union sources that many sub-terra shells had been placed at Yorktown as well as on the Williamsburg Road.[45]

Longstreet was the only Confederate officer to publicly state that he did not support the use of landmines. That statement came in the aftermath of the Yorktown incident, but by the end of the war, Longstreet also had converted to their use to protect the long Confederate line at Richmond and Petersburg.

In addition to Rains, a handful of other ex-Confederates tried to spread torpedo knowledge after the Civil War. Thomas E. Courtenay, developer of the coal torpedo, traveled to Canada and to London, where he made efforts to interest the British government in purchasing the device. James H. Tomb went to Brazil, where he served as a technical expert for the Brazilian military in antiship mines during the War of the Triple Alliance (1864–1870).[46]

While most Confederates had supported the use of landmines during the war and continued to think during the postwar years that it had been justified, Northerners uniformly condemned their use during and after the conflict. In their memoirs and unit histories, Union veterans continued to refer to torpedo incidents as outside the pale of justified warfare. John W. Burke of the Eighty-first New York thought anger created by the Yorktown incident led his comrades to fight harder at the battle of Williamsburg on May 5, 1862. He called the booby-trapping of the Yorktown streets examples of "individual murder." William Simmons, a former naval officer, expressed this theme very well. "There is about the torpedo a peculiar element of uncertainty and deception, like the midnight assassin striking without warning, and being generally encountered when and where least expected."[47]

But no matter what the feelings of average Northerners about torpedoes, the Federal government recognized the necessity of starting a torpedo program of its own. As we have seen, it began to do so by 1864 in a limited way, focusing on antiship mines, railroad explosives, and bridge destruction rather than devices designed to directly kill men. None of these devices were used widely, and none appear to have been effective in destroying the intended target.

After the close of the war, however, the army and navy began a concentrated process of studying Confederate torpedoes in view of starting a long-term program of development and use. Engineer Peter S. Michie, who had served in the Army of the James during the Petersburg campaign, wrote a short study called "Notes Explaining Rebel Torpedoes and Ordnance" while stationed in occupied Richmond during 1865. He accurately described the Rains landmines deployed at Petersburg and the electrical detonation system used at Fort Fisher, identifying the generator as a Wheatstone machine. He also detailed the horological detonation device used in the terrible City Point explosion. His notes remained unpublished until 2011.[48]

The next army study of Confederate torpedoes immediately followed Michie's short document. Capt. W. R. King, who had been involved in the development of the Federal antiship mines in North Carolina, quickly put together a book on torpedoes for the use of his fellow officers in the Corps of Engineers. Published in 1866, it was the first book of its kind to appear in the United States, probably in the world, and had been prepared under the direction of chief engineer Richard Delafield. King took his subject seriously, terming torpedoes a "great auxiliary in modern warfare." He hoped his study would bring up torpedo technology "to a degree of perfection commensurate with its importance."[49]

In his *Torpedoes: Their Invention and Use*, King advocated their employment in defense of fortifications but accurately concluded that their technical development thus far limited their effectiveness in operations. He thought landmines were "more defective than almost any other apparatus for war purposes." It was difficult to time their detonation for the most opportune moment, and they inevitably produced only limited damage. "Aside from the *moral effect*, they were nearly as dangerous to friends as foes," King concluded. He recognized that this was due mostly to the infancy of the weapon and that all problems were likely to be worked out with time.[50]

King proposed an improved landmine using a fifteen-inch shell with a scored casing so it would fragment more uniformly. It could be fitted with a contact fuse or set off by electricity and, to avoid repeating the tragedy of Batchelder's Creek, could be turned on and off to minimize accidents. King also proposed placing a flat stone underneath the shell and logs on the side facing friendly troops to better direct the force of the explosion upward. He advocated the planting of this weapon as near a fort as possible to aid in its defense. "The shells should be placed in quincunx order, their centres about three feet below the surface of the earth, and from twenty to fifty feet apart." What King envisioned was a modernized form of the old fougasse.[51]

But King was more impressed by what he termed offensive land torpedoes than by defensive devices such as his fifteen-inch fougasse. "The bridge torpedo . . . could hardly be improved excepting perhaps, by the substitution of gun-cotton for the powder." Guncotton consisted of cellulose taken from cotton, treated with concentrated sulfuric acid and nitric acid to produce cellulose trinitrate. It had six times the generative power of gunpowder and produced less smoke. While initially developed in Europe by 1846, a safe way to process it was not worked out by Frederick Augustus Abel until 1865. King expected that once guncotton became more generally used, it would be applied to landmines to increase the power of the explosion.[52]

King expressed a fairly accurate assessment of the state of torpedo technology as it stood in 1866. But he discounted the technical effectiveness of contact detonation devices as developed by Rains and overemphasized the importance of electrical detonation. Despite his assertion that landmines were an important—if future—element in military operations, King's vision of them was rather hidebound by tradition. His own conception was heavily dependent on the ways the old fougasse had been deployed rather than influenced by the forward thinking of Rains's torpedo belts, about which King apparently knew nothing.

Historian Norman Youngblood accurately characterized King's book as "far from a resounding endorsement" of torpedo warfare.[53] But the book does offer us a marker in the developing conceptions of landmine use among Northern officers. They were some distance behind Rains in their understanding of the tactical and technical potential of the new weapon, but they were slowly catching up.

The navy also was trying to come to grips with its experience of torpedoes during the war. Lt. Commander John S. Barnes authored a book entitled *Submarine Warfare*, which was published in 1869. He noted that ever since the Federals discovered the first Confederate underwater mine a short distance up the Savannah River in February 1862, they had a good deal of difficulty with them. Especially where the Rebels planted them by the hundreds to obstruct entry into seaports, the US Navy was "materially checked." Barnes admitted that the underwater mine represented "the most destructive and terrible" application of gunpowder to warfare yet devised.[54]

Despite the moral arguments raised by Confederate use of these engines of war, the rebel government persisted in employing them. Barnes thought they excused their own conduct by citing the superiority of Northern naval power, basing their justification on "their feebleness and necessity." Persistent use by the enemy forced the North to gradually adopt the same weapon. Underwater mines had become by the immediate postwar era "an acknowledged instrument of war" in most advanced nations.[55]

Both the army and the navy committed their resources to torpedo warfare in 1869. The Corps of Engineers gathered the means to construct systems of landmines set off by electricity, and the navy established a Naval Torpedo Station at Newport, Rhode Island, that year. By 1874, the navy decided to rely exclusively on electrical detonation. Four years later, the school at this station was instructing twenty cadets every year in torpedo warfare.[56]

In an article published in 1878, David Dixon Porter admitted that the US Navy had shown during the war "either a want of intelligence in not using torpedoes, or an excess of humanity, and a rash confidence of easily overcoming a vigilant and energetic foe." But it had learned its lesson and now paid "particular attention to the subject, and at present are as well informed in all that relates to the torpedo, and as ready to discard our false notions of humanity, as any other nation." Porter's article represents another marker in the acceptance of torpedo warfare among high-ranking officers of the military establishment. He referred to the "false sentimentality" of the early war years that had condemned the use of torpedoes "as an inhuman and unchristian means of destroying an enemy." That sentimentality had never prevented anyone from killing an enemy through other means such as artillery fire. "Self-defense is the first consideration with nations as with individuals," Porter continued, and thus, all scruples were fast vanishing "before the necessities of the time." Porter even ascribed to Rains's idea that the torpedo was such a terrible weapon it would avert a decision for warfare among aggressive nations. "It may seem a strange thesis to maintain, that the torpedo is a beneficent invention, yet all peace-loving men should approve of it."[57]

Acceptance of the torpedo proceeded quickly after the Civil War, but it exceeded man's ability to curb the unintended damage caused by these weapons. Many landmines planted by the Confederates remained in situ long after the war ended, and the government made no effort to sweep them up. Clearance tended to be guided by short-term issues, moving the devices out of the way of whatever needed to be done immediately. Especially in the case of Mobile, no effort seems to have been made to conduct a thorough search over all ground where the devices might have been laid. And even at other places, it took some time for the army to clear them from obvious places as well.

The torpedo belt near Fort Harrison along the Richmond-Petersburg line was still intact as of September 1865, when journalist John T. Trowbridge visited the area. They were in a compact and easily seen spot and thus posed less danger to unwary visitors. Even so, a civilian living nearby warned Trowbridge of the mines and told him how to avoid them.[58]

When writing the history of his regiment, the Thirty-third Iowa, A. F. Sperry mused on what might happen in the future when landowners near Mobile unexpectedly found a torpedo. Despite efforts to clear them from Spanish Fort and Fort Blakely, "it is probable that many of them were overlooked. Likely enough, the farmer who shall yet plow over the land, where the outer lines of Spanish Fort and Blakely extended, may suddenly strike the cap of one of these terrible shells, and be blown to pieces. Thus the weapon the father prepared, may turn against the son."[59]

Bvt. Capt. Charles J. Allen blamed the Rebels for this continuing problem. They had been careless in marking the plantings at Mobile, scattering them quickly without proper records. "Many of them could not be found," he reported in October 1865, "and are liable at any time to injure persons, who from curiosity, or other motive, may visit the ground."[60]

Information about battlefield reclamation is difficult to find, and thus, we have no proof that civilians were killed by the Mobile mines long after the war. But relic hunters discovered five of these landmines near Mobile in 1960. They were twenty-four- and thirty-two-pound spherical shells, armed with Rains's sensitive fuses. The gunpowder was reportedly in good shape, and thus, the mines were still dangerous after nearly a century in the ground. Another Mobile landmine was discovered at a different time, consisting of a ten-inch mortar shell with a safety cap on the fuse. It was found near Fort Blakely on a logging road frequently used by trucks but had never gone off before erosion exposed the dangerous device.[61]

The navy was more careful to search for underwater mines in Mobile Bay than the army was in searching for landmines near the two Confederate forts. This was prompted by the need to clear the waterways for commerce. The Federals found many underwater mines but could not locate all of them. As of October 1865, Allen reported that several transports had hit mines and been destroyed or damaged.[62]

This was the most pernicious legacy of torpedoes during the Civil War. No matter how angry Northerners became in the aftermath of a landmine incident during the conflict, at least there was a countervailing argument that wartime necessity supported their use. There was no justification for the danger these infernal machines continued to pose after the war was over. Killing and maiming innocent civilians was a human tragedy that no one who advocated the use of landmines seemed able to avert.

Chapter 12

Global Developments after the Civil War

"The Effective Stigmatization of the Weapon"

The Crimean War combined with the Civil War initiated a new phase in the history of antipersonnel weapons planted in the ground. The era of the fougasse was over, and the era of the modern landmine had arrived. While the Nobel mines at Sebastopol had been designed and manufactured expressly for the purpose, the Rains mine in the Civil War had been improvised from artillery shells. The trend was definitely toward the use of improvised mines from the 1860s through the various wars around the world that were to come, including World War I. Only during the 1930s would the era of the improvised landmine end and the advent of specially designed and manufactured devices finally settle in.

Meanwhile, the terminology associated with landmines began to change soon after the end of the Civil War. The term *torpedo* continued to be used for landmines after Appomattox, but it soon began to be applied only to a new naval weapon. In 1866 Briton Robert Whitehead developed the first self-propelled device that we today call a torpedo. By the 1890s, *torpedo* ceased being used to refer to landmines and was applied exclusively to Whitehead's invention. This was logical considering that the term *torpedo*, as explained in the preface, had originally been used to refer to a fish. The naval torpedo ran on compressed air and used guncotton as the explosive. The first successful use of it occurred during the Russo-Turkish War of 1877–1878.[1]

Landmines were used in many conflicts after 1865, and the pattern of use generally followed what we have seen in the Civil War. During the War of the Pacific, 1879–1884, the Peruvian army utilized mines in at least two instances. Operatives planted a number of them to protect the land approaches to the port city of Arica. Consisting of one to thirty pounds of gunpowder, some were activated by pressure, while others were detonated electrically. When Chilean soldiers tangled with this system in June 1880, three were injured, but other Chileans captured the Peruvian engineer responsible for planting them and compelled him to produce a map showing their location. On another part of the defense line protecting Arica, the landmines failed to stop a Chilean assault that captured the city.[2]

When Chilean forces approached the Peruvian capital of Lima, they ran into more landmines. The outlying town of Chorrillos was protected by mines and booby traps. Dogs set some of them off, and a handful of Chilean soldiers were also injured by them. This so angered the Chileans that they killed a number of captured Peruvians after the surrender of Chorrillos on January 13, 1881, and Lima fell soon after.[3]

The British embraced the use of landmines in several colonial conflicts. They used them in the Zulu War of 1879. In the Sudan campaign of 1884, they employed landmines to aid in the defense of posts. The British deployed both pressure-sensitive and electrically detonated mines for this purpose. They also used them in the Second Boer War of 1899–1902 to protect bridges and railroads, laying down sixty mines at Ladysmith that were rigged with a system of electrical detonation.[4]

The pace of landmine use picked up during the Russo-Japanese War of 1904–1905. Mostly acting on the defensive in this Manchurian conflict, the Russians made extensive use of landmines. They tended to place them in several rows fifty to two hundred yards before their fortifications and used up to seventy pounds of gunpowder in each charge. The Russians knew that the physical damage to Japanese soldiers was minimal but counted on their moral effect. German field regulations of the era embraced the use of this type of weapon but recommended "automatic mines" so as to reduce the time and effort needed to supervise the mine belts. The Russians in Manchuria could not decide if that was the best way to set them off. They found that contact mines often proved fatal to friendly troops but agreed with the Germans that electrically detonated mines required too much care and attention. Thus, they used both types in the conflict.[5]

While European armies found little wrong with landmine use other than technical and tactical issues, the American army tried to limit their use on moral grounds. The *Engineer Field Manual* of 1909 "set limits on the acceptable and legitimate use of landmines," in the words of historian Norman Youngblood. It advocated their use as an aid to the defense of a fortification, but not in areas where enemy troops could move or assemble. The idea that mining could be conducted "only in areas obviously prepared for defense held sway in the American military through World War I." This was a holdover from the intense moral revulsion against torpedo warfare as practiced by the Confederates, and it lasted half a century.[6]

In this way, the United States was, for a time, ahead of the international community. During the First International Peace Conference at The Hague in 1899 and during the second conference of 1907, widespread discussion about the use of underwater mines took place, but no discussion about the regulation of landmines was heard.[7]

World War I, with its trench stalemate along a western front that stretched for hundreds of miles, seemed to be the perfect venue for widespread use of landmines. But that did not happen. The old pattern set by the Civil War still prevailed. Landmines were used sporadically, and they were improvised from artillery shells and other munitions, overwhelmingly employing contact detonation rather than electrical devices. The Australians placed a small number of mines during the Gallipoli campaign, using fifty pounds of a new high explosive called ammonal as the charge and setting it off with a slab of guncotton activated by pressure. This was done at the end of the campaign, in December 1915, as part of the Australian plan to evacuate the peninsula.[8]

There were relatively few instances of landmine use during World War I on any front. When the British introduced the tank in 1916, it inspired engineers to develop antitank mines. Again they improvised them from artillery ordnance. But the most common form of landmine eventually used during the war of 1914–1918 was the booby trap. The Germans excelled in this particularly hideous form of landmine warfare, planting thousands of hidden explosive devices in areas they planned to evacuate during the latter stages of the conflict. In an effort to straighten out their front and make a more compact position, the Germans pulled out of a large swath of territory and fell back to the Hindenburg Line in 1917. They planted contact mines in buildings and along roads, especially crossroads, that were likely to be used by the British. Some of these were time-delayed bombs that exploded without warning hours, days, and even months later.[9]

During the last months of the war on the western front, the Germans increased their planting of landmines and booby traps as they conducted a fighting retreat. From August 5 to November 11, 1918, British engineers removed 6,714 landmines from evacuated ground to clear the way for Allied movement and occupation. They also found 315 delay-action mines and 536 other types of booby traps. When the Germans agreed to an armistice to take effect on November 11, 1918, they also agreed to provide information about the delay-action mines to the Allies. The longest delay device was set to go off January 11, 1919.[10]

World War I represented the end of the period of landmine use that had begun with the Crimean and the Civil War. During that period, however, the two primary uses of mines had been to aid in the defense of a fortified position and to use as a surprise element (a booby trap) in areas where opposing troops might move, assemble, or quarter. The western front saw the greatest fortified position in world history, and yet, there were no appreciable number of landmines deployed to aid in its defense. But the extensive use of booby traps was notable. Perhaps other defensive devices, such as the machine gun, extensive barbed-wire entanglements, and masses of heavy artillery, inhibited the use of landmines as an aid to defense along the western front. No one developed a standardized landmine for mass production and deployment, probably considering that step unnecessary with all the new firepower that was available.[11]

As far as the American army was concerned, World War I finally ended moral concern about landmine use. In a manual on field fortification issued in 1928, one finds that the US Army dropped its attempt to limit mine use to the defense of fortified posts and authorized their deployment in roads and places where the enemy might assemble. The Civil War generation's outrage about torpedoes had run its course, and the American army's attitude toward landmines was now in line with that of other armies around the world.[12]

In the immediate aftermath of the Great War, the lessons learned about landmines seemed to be that they were useful only as booby traps. The British manual of field works, issued in 1921, stressed them as impediments to enemy movement after one gave up territory and did not even mention them as an aid to defending a position. "There is ample field for ingenuity and cunning in constructing these devices," the manual continued. But it also recommended a carefully thought-out plan to deploy them and the maintenance of detailed records so they could be taken up again. Landmines had still not gained the attention of those who sought to limit the horrors of warfare through international agreements. Even though such agreements dealt with the use of poison gas, bacteriological weapons, the treatment of prisoners of war, and the treatment of civilians, they paid no attention to the use of landmines.[13]

A new era in the history of these devices began by the early 1930s, as several nations developed standardized types of landmines that could be mass-produced for deployment by millions in the field. Germany led the way. Even before the advent of Adolf Hitler, German technicians developed the Tellermine in 1929, followed by the S-35 Schutzen mine of 1935, which sprang out of the ground before exploding. The Soviet Union also developed a standardized landmine in 1935. The other nations soon to become involved in World War II were behind the curve, compared to Germany and the Soviet Union.[14]

Those competing nations soon found they had to catch up. World War II witnessed the full deployment of the landmine as a modern weapon, placing it squarely within the operational mode of all nations as a factor in military operations. As several historians have put it, the landmine reached "full maturity" with the global war of the 1940s. The conflict spurred further development. While the Germans had only one antipersonnel and two antitank mines in 1939, they developed ten kinds of antipersonnel mines and sixteen types of antitank mines by 1945. It has been estimated that German forces laid as many as thirty-five million mines during World War II. The Soviets developed sixty-one standard types of landmines by war's end, including twenty-six different fuses that included activation by pressure, by pulling, by electronic frequency, and by vibration. Russian forces deployed up to sixty-seven million landmines during World War II.[15]

The United States began the war some distance behind its chief enemies but advanced as the conflict lengthened. The Americans developed their first antitank mine and their first bouncing antipersonnel mine in 1941. Two years later, they made a nonmetallic device and developed ten standard mines by 1944.[16]

Along with this enormous expansion of deployment came a sophisticated system of doctrine and method that generally followed the outline of Gabriel Rains's thinking during the Civil War. This is not to imply that Rains or the Civil War influenced the Germans or Soviets, but it does indicate similar lines of doctrine concerning this weapon that, in its tactical implications, did not change much over the decades.[17]

In terms of morality, the arguments had ceased by the 1940s—landmines were here to stay. In terms of tactics, the same basics applied. In terms of technology, the landmine had progressed very far beyond the relatively primitive, makeshift methods of the Civil War era.

The technical aspects were impressive. In addition to the bouncing mine, which elevated three feet above the ground to spray a wider area with fragments, a wood-cased mine was developed to make it more difficult for enemy troops to find the buried device in clearance operations. The Germans also developed mines encased in ceramics and concrete for the same purpose. The Japanese developed a more limited number and type of landmines and even relied on a chemical contact mine for deployment on contested beaches. They also improvised mines from artillery shells and unexploded Allied bombs near the end of the conflict. The Italians developed a landmine that could be scattered from airplanes. Both the Germans and the Japanese used coal torpedoes for sabotage purposes during World War II.[18]

The Germans created a full-spectrum approach to the tactical deployment of landmines. This can be divided into four main categories—in front of a heavily defended position, as a nuisance factor along lines of communications, randomly spread through areas likely to be occupied by the enemy, or as dummy fields (areas apparently mined to confuse and deceive the opponent). They usually planted the mines in patterns to create a wide and deep area of danger. Covering the field with defensive fire greatly increased the tactical value of the landmines. Fighting a defensive war by the latter part of the conflict, the German army developed very sophisticated methods of creating and defending minefields.[19]

The Americans found themselves at a disadvantage when they initially began to deal with German minefields because their training and doctrine lagged behind that of their enemy. The psychological effect was still considered the most important factor in landmine warfare, especially when dealing with booby traps, which the Germans also planted widely in evacuated areas. When the Americans actually deployed minefields—as to defend their beachhead at Anzio—they tended to do so without much order, which led to friendly casualties. Actual losses due to landmines and booby traps combined tended to be not inconsiderable. In the northwest European campaign from June 1944 to the end of the war, a bit more than 20 percent of British and American tank losses were due to landmines. In contrast, German tanks accounted for slightly more than 14 percent of Allied tank losses. Landmine casualties among infantrymen were much lower than among tanks. In the American army in Europe, 4.4 percent of manpower casualties were caused by landmines.[20]

The British, who also were behind the curve in landmine warfare, reacted quickly to take the lead in developing methods of detecting and clearing enemy fields. The Americans also contributed to this tactical effort by using flails at the front of tanks to "beat paths through minefields." In the end the only real way to deal with a minefield in a full and permanent way was to carefully find and dig up each device. This proved to be a dangerous job. The Tenth Engineer Combat Battalion lost fifteen men who were killed and forty-three who were wounded while clearing twenty thousand landmines north of Naples during the Italian campaign.[21]

The use of landmines increased as World War II lengthened. The British deployed 350,000 of them in southeast England after the fall of France to defend the island against an anticipated German invasion. In North Africa, mass use of landmines progressed to the point where the weapon now became a repeated feature on the battlefield. The Russians used half a million landmines to defend the Kursk Salient in 1943, but estimates as to the total number of landmines used by all belligerents during the Second World War vary widely. Conservative estimates place the number at no more than one million, while extreme estimates place it at more than one hundred million.[22]

Even if one accepts only the lowest estimate, there is no doubt that World War II saw the deployment of far more landmines than had ever been used by all nations in world history before 1939. This presented a gargantuan problem of clearance. As soon as the war ended, the Allies used German prisoners to help demobilize minefields. Several thousand prisoners participated, and 8 percent to 17.5 percent of them were killed or wounded during 1945–1946 by their own mines. The Allies pushed this effort as quickly as possible because negotiators at the Geneva Convention were likely to prohibit the use of prisoners for mine clearance. The Convention delayed this decision until August 1949 to give the authorities more time to take up the majority of these devices. After clearing a field, the prisoners often were marched across it to demonstrate to local civilians that the area was clean.[23]

Although some sources claim that ninety million landmines had been taken up in two years of de-mining, many others remained in place for years to come and caused civilian casualties. Far less was done in North Africa than on the continent of Europe, and much land was still not usable in Libya and other places along the southern rim of the Mediterranean Sea as of 1980. By 1981, 4,094 civilians had been killed and 8,774 wounded by landmines in Poland alone.[24]

Landmines and booby traps, horrific to the individuals who encountered them, still paled in comparison to the other horrors of World War II. Yet that conflict ushered in the industrial age of landmine warfare. The most important influence of the conflict on the future was the development of technically sophisticated, standardly produced devices that could be deployed by the millions in easy and dangerous ways. Regional conflicts conducted under the umbrella of Cold War issues became the venue for how the future of landmine warfare played out after 1945.

Development of improved mines continued during the Cold War years as it had during World War II. The Americans devised at least a dozen new landmines from 1945 to 1962. One of them was a directional mine, which could aim projectiles toward the person who tripped it rather than spraying them in a standard pattern. Another was a throwback to Nobel's Sebastopol mines in that it was based on chemical reaction. The most effective American landmine of all time was the M18 Claymore. The A1 version of the Claymore contained 1,600 steel balls propelled by one and a half pounds of high explosives inside a plastic case. Set off electronically, it sprayed the balls up to fifty yards away.[25]

By 1972 the British developed mechanical mine-laying capabilities to more efficiently create mine fields. The most useful technology, however, lay in the area of self-neutralizing concepts. The French developed a mine that would disarm itself after thirty-one days in the ground, by 1974. The Swedes went a step farther and developed, by 1982, a mine which could be armed or disarmed at preset times. Thus, it could go on or off within a span of 30 to 180 days. This technology was a brilliant solution to the long-range problem of mine clearance and neutralized the danger to civilian populations over time.[26]

But the problem lay in the tendency of belligerents to use any mines available rather than the latest self-neutralizing technology. It has been estimated that by the 1990s, more than six hundred different kinds of landmines had been developed worldwide, offering everyone a wide choice of weapons.

Landmines became increasingly prevalent in guerrilla conflicts during the Cold War era. They were relatively cheap weapons for a weak force to obtain as well as a natural ally to counter-insurgency operations. From 16 to 30 percent of all American soldier deaths in Vietnam were caused by mines. In some units the percentage was much higher. The First Marine Division lost 57 percent of its killed and wounded to landmines during the last half of 1968 alone.[27]

Mines continued to play a significant role in conventional military operations during the Cold War. In 1965 the Americans planted a heavy belt along Korea's thirty-eighth parallel as a deterrent to the North. The Iraqis planted up to two million mines during the First Persian Gulf War of 1990–1991. Coalition forces, led by the Americans, employed similar tactics to breach these fields as was common during World War II. Dealing with the mines after the conflict proved to be more deadly than breaching the field under fire. Within six months after the war ended, 70 Coalition soldiers were killed and 360 were wounded due to the mines and unexploded ordnance combined. In addition, 1,700 civilians were killed by the same cause by 1993.[28]

Yet it has been in guerrilla conflicts during the Cold War era that the landmine has been most extensively used in history. One historian has called it "the main weapon of war" in these conflicts. The aftereffect of their use was traumatic for society. "Mines affect almost every aspect of life in countries emerging from conflicts," historian and de-miner Mike Croll has written. "They reduce agricultural production, prevent refugee return, hinder reconstruction, make roads impassable, cause large numbers of casualties and perpetuate the fear and anxiety caused by war." These problems were accentuated when they occurred in developing countries that had many unsolved social problems even before the conflict started. In many places, local governments proved incapable of funding or managing clearance programs, and outside sources of funds and expertise were required. Wealthier governments, the United Nations, and private charitable organizations have contributed. After the Soviet pullout from Afghanistan in 1987, international concern about these mine-clearance problems gathered strength until a range of options for impoverished countries faced with the need to deal with millions of leftover mines became available.[29]

The Cold War era finally brought the issue of landmine use to the attention of international negotiators concerned with limiting the use of certain weapons in warfare. The Geneva Convention of August 12, 1949, implicitly acknowledged landmines as legitimate by forbidding the use of prisoners of war to clear them. The Convention on Prohibition or Restrictions on the Use of Certain Conventional Weapons, a United Nations document signed by forty-one countries in the 1980s, required that landmines be marked and their location recorded for easier clearance after the conflict. But it addressed only wars between states, not within them, even though the latter saw more landmine use than the former.[30]

Efforts to ban landmines began to take shape at about the same time. In 1981 the Inhumane Weapons Convention included booby traps and antipersonnel weapons, but it failed to gain widespread support. A few years later, the International Campaign for a Landmine Ban sponsored a conference held at Ottawa, Canada, in October 1996. The result was the Antipersonnel Mine Ban Treaty signed on December 3, 1997, at Ottawa. Officially termed the Convention on the Prohibition of the Use, Stockpiling, Production and Transfer of Anti-Personnel Mines and on their Destruction, it became effective on March 1, 1999, but as of that date, only 65 of the 133 signers had ratified the agreement. It prohibits the use and sale of these devices and obligates clearance as well as care for victims. The legal basis of this treaty is a recognition that landmine use does not differentiate between military and civilian targets.[31]

The Antipersonnel Mine Ban Treaty has been very controversial. The major nations with landmine capabilities, including Russia, China, the United States, and India, have refused to sign it. Their argument has been that the real problem is irresponsible use by irregular forces rather than by modern armies. These nations prefer tighter control over the dispersal of landmines rather than a total ban on them. Some commentators predicted that the ban would not work and stressed the careful policy worked out by the United States to govern potential deployment of antipersonnel mines as the best course of action for all nations to pursue.[32]

As of 1999, the global problem presented by antipersonnel mines had grown to gigantic proportions. Even if one discounts the inflated statistics promoted by the more zealous promoters of mine banning, the numbers are still appalling. According to Andrew C. S. Efaw, decades of landmine use has impregnated the earth with deadly devices that continue to kill and maim. The most reliable estimate is that up to fifty million landmines still infested populated regions around the world by 1999. The same estimate places the number of deaths and injuries caused by them during the last two decades of the twentieth century at one million. Life for the professional de-miners is no less dangerous. In France alone, 630 of them lost their lives from 1945 to 1999, and the problem was still growing at the turn of the twenty-first century. According to reports, an additional two million new mines were being planted each year, while only eighty thousand were being taken up. It was estimated that up to one hundred million landmines had been manufactured and were stored in stockpiles in many different nations by this time.[33]

But with the passage of time, there were signs that the Antipersonnel Mine Ban Treaty of 1997 began to have a watershed effect on the problem. First, in redefining the legal question, it has separated landmines from aerial bombs and other ordnance that also threaten civilians and places them in a category of their own as violating international principles. As such, the treaty represents "a fundamental re-conception of the meaning and limits of military utility," according to legal scholar Adam Bower. It was "a dramatic change in the international status quo regarding the legitimate conduct of warfare."[34]

Bower also saw signs that the ban was actually working to reduce landmine use. Statistics indicated that the production of mines was declining both in states that did not sign the treaty as well as in states that did sign it. While fifty nations had landmine production capacity before 1997, that number had declined to only twelve by 2015. Thirty-four nations had exported landmines before 1997, but by 2015, there were no confirmed instances of exportation anymore. Bower credited "the effective stigmatization of the weapon initiated by the mine ban movement and treaty" for these important changes. Hopefully, he is correct in asserting that "the use of AP mines has now become an aberration in international practice."[35]

If the signs described by Bower are indicative of the future, the trend toward extensive use of antipersonnel mines, which Gabriel Rains promoted in the Civil War, has finally run its course by the early twenty-first century. The effort to clean up after this epoch continues apace. Leftover landmines are often folded into the more general problem of unexploded ordnance and called Explosive Remnants of War, or ERW. The *Journal of ERW and Mine Action*, published by the Center for International Stabilization and Recovery at James Madison University, highlights this global effort with a mix of historical perspective and modern-day processes, problems, and successes in the removal of dangerous ordnance.[36]

Conclusion

The Civil War was a watershed in the development of landmines. It saw the first large- scale deployment of the device at several locations rather than just one, as had earlier happened at Sebastopol. The most effective detonation system, Rains's sensitive primer, and the first doctrine concerning how to use the device in its tactical setting appeared during the Civil War. This doctrine, also developed by Rains (even though not published), essentially captured the main points of how to use the landmine for the rest of its existence. Rains's doctrine did not appear in print until recently, so it did not influence landmine use after the war. Other operatives in many other countries developed their own doctrine along the lines conceived by Rains because those lines could be worked out through experience with using the device.

The significance of the Civil War in the overall history of landmines has been muted only because previous historians have not conducted the kind of deep research necessary to uncover that significance. Until Herbert M. Schiller edited Rains's "Torpedo Book," for example, most historians had no idea that the eccentric general had spelled out the outlines of torpedo warfare with such a high degree of accuracy. Other historians have hinted at the significance of Civil War landmines but have not developed that significance for modern readers. Or, in other cases, some historians have given Rains too much credit; after all, his inflated hopes for the tactical effectiveness of torpedoes proved to be completely wrong, although he never admitted that failure.[1]

Rains played a large role in the story of Civil War landmines, but it has to be remembered that several other men played roles that were important as well. Rains gained more notoriety because he was a consummate self-promoter after being "outed" in the aftermath of Yorktown. As indicated in previous chapters, Rains was shrewd when it came to developing the technology and tactical deployment of his weapons but quite off the mark when predicting their impact on military operations.

In summing up the three threads in the story of Civil War landmines, the morality, tactics, and technology of the device, it is necessary to begin with an estimate of how many land torpedoes were planted during the conflict. Landmine historian Mike Croll has estimated the total of land and underwater mines used in the Civil War at under 20,000, and he undoubtedly is correct.[2] My own counting of landmines, based on reliable evidence, is that at least 4,000 sub-terra devices were planted by the Confederates at a minimum of fifteen locations during the war. At least 163 of those torpedoes were exploded by victims, a small number of whom were friendly Confederates and civilians. This means that only 4 percent of the total number of landmines that were planted resulted in a successful attack on someone.

When Croll noted the significance of Rains's pressure-operated landmine, he concluded that "it is reasonable to credit (or reproach) the Americans with the development of the first operational devices" in torpedo warfare.[3] The ambivalence in his phrasing is noteworthy. While Rains never felt any remorse about developing a weapon that killed without warning and harmed civilians, friends, and enemies alike, many Northerners were livid with anger over this development. They were primed to do so by the torpedo crime news of the 1850s, when similar devices were used in murder attempts.

It is not surprising that most Northerners reacted to the planting of landmines by branding the act as a crime, out of the pale of civilized warfare. Some Northerners differentiated between a legitimate use of the new torpedo (planting before a fortification as an aid to its defense) and the spreading of these deadly devices in areas to be frequented by Union soldiers (booby traps). The latter was far more heinous in their view than the former. But many other Northerners did not differentiate between legitimate and illegitimate uses of landmines. They branded all torpedoes as barbarous.

Sherman represents much of the emotional reaction to torpedoes. He became angry at their use more than once during the war. When contemplating the issue calmly, he differentiated between legitimate and non-legitimate use. But when faced with the horrible sight of their impact on his men, as at Pooler Station, his anger overrode calm reflection, and he branded them all as barbaric.

McClellan began the process of using prisoners to dig them up, and subsequent officers followed the natural tendency to expose representatives of the army that had planted the devices rather than risk their own men. Sherman and Steele were prominent in this regard. The Confederate prisoners of course did not like this, but most of them submitted rather than risk a prick from a Federal bayonet or being shoved onto suspicious ground by force to see if they would step on a torpedo. One must give the Confederate authorities credit for refusing to use Union prisoners in this way when called on to do so by Beauregard. But then, their men were never victims of Union landmines, and they did not know the visceral feelings of anger and rage that these devices generated when one saw his comrades blown to pieces.

Confederate reaction to the use of landmines was a bit mixed. Given their desperate war effort, most people who commented supported the use of landmines as a force multiplier. Rains easily convinced Davis and the Confederate government to adopt them as official policy, becoming the second government next to the Russian imperial authority to support a landmine capability.

Confederate prisoners at Mobile expressed fear of retribution because of the widespread use of torpedoes. More than one Union soldier reported that captured Rebels expressed disgust with their use. This represented a silent criticism of their government's policy, but it probably was not very widespread.

We do not know for sure if America was unique in this moral argument. More research is required to find out if the British public reacted negatively to the Russian use of torpedoes at Sebastopol. But it is possible that the moral argument was unique to the United States, probably because of the torpedo crime news of the 1850s.

We do know that the moral argument faded in the United States with the passing of the Civil War generation. It disappeared by the turn of the century, when landmines were used in the Russo-Japanese War of 1904–1905 and in World War I. By this time high explosives were turning artillery into the most deadly weapon on the battlefield, and the technology of landmines did not advance until the 1930s. The German use of poison gas in 1915 sparked a moral revulsion in Britain that resembled the public outcry against Confederate torpedoes among Northerners fifty years before. Characterizing gas as an "atrocious method of warfare" and a "diabolical contrivance," calls arose for retaliation against this "dastardly and uncivilized weapon."[4]

The Civil War figure more responsible than anyone else for developing an effective landmine was Gabriel J. Rains. In the Fort King incident and in the Civil War, he was motivated, at least in part, by revenge. Rains thus automatically assigned to himself the role of the offended reacting to an act of the offender. He never looked back, in a moral sense, firmly convinced he was doing the right thing regardless of criticism. But Rains obviously felt the sting of that criticism and was sensitive to it. That is why he acted slyly at Fort King and at Yorktown, doing his work without authorization from superiors. He also was careful to disassociate himself from most of the torpedoes that had been planted at Yorktown even though all the evidence points to him as the instigator of those plantings. Only after being exposed did Rains lobby for official approval from superiors and readily obtained it from Davis, who became his chief patron for the rest of the war.

Davis also wanted to avoid the possibility of moral outrage over the use of landmines. He maintained a semi-secret status for the program, ostensibly, because he did not want the Federals to obtain information about the "secret" weapon. But evidence also indicates that he wanted to avoid moral criticism too. As indicated by the story concerning Oladowski and the coal torpedo, Davis wanted to maintain a public persona of high-mindedness, claiming he would never resort to sabotage even while he was doing so on the sly. We can assume that the practitioners of torpedo warfare among the Confederates recognized the moralistic questions associated with their work and, to a limited degree at least, tried to avert them.

While Confederate authorities officially adopted a torpedo campaign, despite the moral questions, Northern authorities refrained from deploying antipersonnel mines during the war. But they also failed to officially condemn the use of torpedoes. They had an opportunity to do so in the famous Lieber Code, which was adopted midway through the war. Born in Berlin in 1798, Franz Lieber had served in the Prussian army, where he had been wounded in the last stages of fighting during the Napoleonic wars. He earned a doctoral degree at the University of Jena before immigrating to the United States in 1827, securing a chaired position at Columbia University in New York thirty years later. Lieber proposed putting together a guide to the moral conduct of war in November 1862, and Henry W. Halleck readily agreed, working closely with the German to develop it.[5]

Halleck issued Lieber's document as General Orders No. 100, dated April 24, 1863. It was later published as *Code for the Government of Armies*. Based on legal precedent, the code had a strong tone of humanitarianism, but Lieber did not specifically mention torpedoes even though he referred to other means of exacting death and injury that Northerners felt were unjustified, such as poison. But the most relevant part of the code for landmines is Section 16: "Military necessity does not admit of cruelty—that is, the infliction of suffering for the sake of suffering or for revenge, nor of maiming or wounding except in fight." In Section 28 he drew a sharp line between retaliation and revenge, the latter tending to draw conflict toward "the internecine wars of savages," and thus was unacceptable. Rains specifically mentioned both retaliation and revenge as motives for his actions without differentiating between the two concepts.[6]

Ironically, Lieber's code seems to have had little impact on Northern opinion during the remainder of the war. It is hardly mentioned by anyone in official documents or personal accounts. But the code became widely known, respected, and used after the war. It remained the basic code of conduct for the American army for the remainder of the nineteenth century, and a new manual for land warfare adopted in 1914 retained most of its ideas.[7]

Brig. Gen. William F. Barry read General Orders No. 100 and wrote a long letter detailing the Yorktown torpedo incident for Lieber. As McClellan's chief of artillery on the Peninsula, Barry was well placed to report on the landmines at Yorktown. He wrote it was clear that Gabriel J. Rains had been the instigator and that his brother had prepared the devices. Many Federal officers recalled Rains' Fort King torpedo in the Second Seminole War and had no difficulty believing he was guilty of the same at Yorktown.[8]

Barry indicated that he personally witnessed the death of a telegraph operator who stepped on a landmine in the street, which was planted near a telegraph pole. Whether he referred to D. B. Lathrop or another operator is uncertain, for various stories about exactly where and how Lathrop tripped a landmine, whether in a house or on the street, had circulated. Barry also witnessed "the horrible mangling by one of these shells of a cavalryman and his horse outside of the main work upon the Williamsburg road."[9]

It was important to Barry that Lieber understand these devices were not placed where they could aid in the defense of a fortified position during an assault. The old fougasse had been planted in the glacis, which was a legitimate aid to defense of a work. Even planting torpedoes in the bottom of the ditch would be permissible. But Rains planted these devices all over the place, inside works, in the streets of Yorktown, and on common roads. Worst of all were deliberate efforts to trick the Federals by attaching the detonation devices to "articles of common use, and which would be most likely to be picked up." These items included wheelbarrows, pickaxes, and shovels. Barry told Lieber that McClellan used Confederate prisoners to find and dig up the torpedoes.[10]

Barry laced his letter to Lieber with loaded language. Using phrases such as "cruel murder," "dastardly business," and "disgraceful trick," he made sure the professor knew his opinion on the use of landmines at Yorktown. Many other Northerners used these phrases as well; they can be taken as the generally accepted viewpoint of the morality of landmine use during the Civil War among Unionists.[11]

Although he did not say it, Barry obviously was concerned that Lieber had not mentioned torpedoes in the code. The Federals never engaged in retaliating on the Confederates by deploying antipersonnel landmines. They seldom had the opportunity to do so, but there is literally no evidence that anyone felt the urge to plant antipersonnel mines, and no plans were laid to place booby traps in a way to deliberately kill or maim Confederate troops.

The Federals refrained from developing a torpedo program until finally relenting at the midpoint of the war. Even then they created weapons designed to destroy material rather than people. Union engineers and technicians developed underwater mines to act against Confederate warships, bridge torpedoes to bring down spans, and railroad torpedoes to tear up track and engines. Moreover, they did not deploy more than a handful of these weapons. Only after the war ended did the US Army and Navy adopt an official long-term program. They trained cadets at West Point and Annapolis in the basics of torpedo use and maintained a small arsenal of devices. While the navy used underwater mines in various conflicts, the army did not use landmines before

World War II. It was easier to justify these devices when they were designed to wreck material rather than people.

The moralistic questions concerning landmines would not entirely disappear even though they faded when the Civil War generation passed away. The horrible effect of landmines on innocent civilians gave rise to a new sense of moral outrage on an international scale as the pace and intensity of landmine use increased during the Cold War. Technicians now began to experiment with mines that became inactive over time to lessen the possibility of unintended victims tangling with the dangerous leftovers of war. International organizations tried to help impoverished countries deal with these terrible dangers to their citizens, whose goal was to reclaim farmland and get on with their lives without being killed or maimed. What Rains had done so much starting in the 1860s played out in untold suffering for millions of innocent people over the next century. Only in the 1990s, 130 years after Yorktown, did nations finally agree to a treaty limiting the use of landmines globally.

It was comparatively easy to develop torpedo technology and understandable that the North would react negatively to the device, but much work had to be done to fit the new contact landmine into the context of operations during the Civil War. Rains played a large role in developing ideas and testing them in practice and, in the process, developing the first doctrine concerning landmine warfare. He had already started at Fort King with a booby trap designed to hurt the enemy for the sake of hurting him, not to defend a fortification. During the Civil War, he mixed the two concepts at Yorktown, and other Confederates who followed him did the same at other locations. Both tactics were successful at killing and wounding a handful of Federals and their horses.

Rains concentrated on contact mines from the start and never had faith in electrical detonation. He worked out the details of how to plant contact torpedoes in a mass before a fortification, marking them with flags and maintaining passages through the belt for friendly pickets. He also worked out a plan for the quick planting of torpedoes along roads to impede Federal mounted raids. He was aware that his method of deployment posed a threat to friendly troops and to civilians living in the area but could not come up with a solution other than to mark the mines or to post sentries to warn the unwary. He never addressed the long-range problem of leaving landmines in the ground for years after they were needed.

Rains grossly overestimated the tactical effectiveness of his new weapon. While he helped to develop the contact mine into a technically reliable device, he could not control how it affected the Federals. Rains imagined the weapon would so terrorize his enemy as to immobilize their movement. He even predicted it could end wars by its immense power to overwhelm the emotions of soldiers. In all these predictions, he was quickly proven wrong. The Federals who encountered landmines were shocked by them. They developed a wariness and concern about being blown up, but by no means were they terrorized or immobilized by this new threat. In fact, officers and men alike quickly adapted themselves to the presence of landmines in their path. They learned how to cordon off areas filled with mines, posting guards and erecting signs to warn their comrades. They experimented with methods to neutralize them without exploding. Even though they used captured Confederates at several locations, there is ample proof that Federal troops also participated in the clearing of mine fields. Everywhere used, torpedoes became an additional hazard on the battlefield but failed to evolve into a campaign-winning device. The Federals quickly learned how to deal with them.

It is true that if landmines had been used on a much more extensive scale during the Civil War, they might have had a bigger effect on tactics. But one must keep in mind that even when standard types of landmines were deployed by the millions during World War II, they also merely caused changes in enemy operations rather than stopping those operations altogether. Armies developed methods of clearing paths through minefields, changed plans so as to advance along different routes, and continued offensive operations despite the large-scale hazard. The deployment of millions of landmines during the Cold War–era conflicts also failed to end war through their ability to terrorize potential victims. In fact, the primary victims of Cold War landmines were innocent civilians rather than military personnel.

Developing effective technology to make a modern landmine was easy. Rains started it by the simple expedient of taking an artillery shell and a percussion fuse and rigging them up in the Florida wilderness at Fort King. His only real contribution to improved technology was in later altering that fuse to make it more sensitive to slight pressure. Exactly when he did that is not clear, but certainly it was done by early 1863 and very possibly as early as Yorktown. In its simplicity and with the easy availability of shells, Rains developed the most effective landmine yet to be seen in world history. The only previous system (developed by Immanuel Nobel) required expert knowledge and specialized material even though it apparently worked just as well as Rains's device. Rains's invention had the advantage of being able to be deployed by an artillery officer or anyone knowledgeable about ordnance. For the next sixty years, landmines would follow the general characteristics as laid down by Rains. He improvised by using preexisting ordnance. Only when the concept of developing a standardized device that was mass-produced in factories took hold in the 1930s did the world witness a new era in landmine warfare.

It is just as important to recognize the limitations of Rains's influence on landmine development in the Civil War as it to credit him for what he did in that process. Although Rains certainly was the most influential figure in the development of torpedo warfare on land during the conflict, he was not the only man working in the area. A number of much more shadowy figures planted mines at Columbus, Mobile, and other places. They often used devices of Rains's pattern, but not always.

Rains played no role in the development of electricity to detonate a landmine. That process had been worked on by others before the war and during it. While electrical detonation was used to explode underwater mines, it was employed only twice during the war on landmines, at Columbus and Fort Fisher. In both cases the system was not touched off by operatives, and so neither can be counted as a full test. There is no doubt that the technology of electricity was advanced enough by the 1860s to be effective in landmine detonation, but it never developed into the primary way to set them off during or after the Civil War. The contact mine remained the most popular throughout history because it was a self-activated device that did not demand constant attendance after planting. About the only advantage of electricity over contact systems was that it allowed the watchful operative to explode mines when enemy troops were near the device and not in touch with it. Another weakness of electricity was that the wires could be damaged or cut as they were at Fort Fisher by artillery fire.

In the end, despite the enormous effort expended by Rains and other men, the Civil War had little overt impact on the general course of landmine warfare in world history. No European observer studied it. Nevertheless, developments in global history are often concurrent rather than a cause and its effect. In other words, it was natural for many men in various countries to think of adapting artillery shells with a percussion fuse to make a modern version of an old fougasse. When soldiers in World War I did this, they were not aware of Gabriel Rains or of the Civil War. Even so, this demonstrates how tightly American military history tends to be intertwined with Western military history in general. Soldiers facing similar situations and with similar means developed similar solutions to tactical problems whether they were fighting in the Crimea, in the American South, or on the Western Front. Civil War personnel were only the first to work these out in ways that would become standard for decades to come.

Appendix

A Medical Perspective on Landmine Injuries

No one made an attempt to record in full the number of people killed or injured by landmines during the Civil War. There are partial lists for a few incidents, such as the accidental explosion at Batchelder's Creek, the sabotage at City Point, and the Federal assault at Fort McAllister. For the other incidents, one must rely on soldiers mentioning names of torpedo victims in their personal accounts. It is therefore impossible to settle on a thoroughly documented number, but one can approach that goal by compiling the available evidence.

Basing his or her estimate on sources such as these, one can conclude that at least 319 people were killed or injured by landmines during the Civil War. That figure is skewed because 191 of them, or 59.9 percent of the total, were victims of the Batchelder's Creek accident and the City Point sabotage. The rest were killed and wounded in a total of fifteen incidents of landmine use. Seven of the 319, or 2.2 percent, were Confederates. Eighty of the 319, or 25 percent, were civilians. At least two women and three children were among the civilian casualties. Soldiers made up the majority of victims with 239 of them, or 75 percent, among the 319.

A handful of surgeons and observers commented on the nature of landmine injuries. Indeed, even modern historians have contributed to this perspective. Kochan and Wideman have pointed out that these devices generated an intense pressure wave for a short distance, although it rapidly dissipated. Because landmines, especially those activated by contact, were positioned to explode within inches of the victim, they could readily lacerate and dismember because the individual was within that zone of intense pressure. In other words, the pressure wave generally did more damage than the fragments of the sub-terra shell. In contrast, when exploding in the air, the same shell mostly harmed individuals by flinging fragments about rather than by the pressure wave. Kochan and Wideman correctly concluded that mounted men were often protected by their horse or mule when it stepped on a landmine. The animals' large bodies absorbed most of the pressure wave and fragments, saving the man from worse injury.[1]

The experiences of many mounted men whose steeds tripped a landmine support the conclusion that animals absorbed most of the damage. George W. Read of the Tenth Indiana Cavalry saw a staff officer whose horse exploded a torpedo on the advance to Blakely during the Mobile campaign. The horse was "literally turned inside out," while the officer was unhurt. A fragment had passed between his arm and body, tearing his coat. The experience "stunned and bewildered" the officer, who "talked incoherently for a while."[2]

But unmounted men who triggered a torpedo were exposed to its full effects. Fifteenth Corps surgeons caring for the casualties of Hazen's attack on Fort McAllister treated thirteen cases of landmine injuries after the battle. "These were generally severe contusions, occasionally lacerating the soft parts, and in one instance tearing off the limbs."[3]

This conclusion was largely supported by the observations of surgeon Samuel Weissell Gross. He was the son of Samuel David Gross, probably the most famous surgeon in Civil War–era America. The elder Gross was the subject of Thomas Eakins's painting *The Gross Clinic*, finished in 1875. The younger Gross was born in 1837 and received his medical degree from Jefferson Medical College twenty years later. He was a surgeon at Howard Hospital, Philadelphia Hospital, and the Hospital of Jefferson Medical College. The younger Gross was noted for his use of antiseptic surgery and radical surgery in the case of cancer patients.[4]

Samuel Weissell Gross concluded from his observations of torpedo injuries at Yorktown and Fort Wagner that the most frequent result was a "complicated fracture" and that "a portion of a limb is carried way." But now and then, "very grave and deeply seated injuries, without lesion of the integument [skin], are produced." Lesions produced by landmines varied from case to case. In addition, he saw one Yorktown victim who had "the entire posterior surface of the left leg and thigh" burned quite badly, "the recovery being attended with contraction of the limb."[5]

In an article published in the *American Journal of Medical Sciences* a year after Appomattox, Gross detailed five case studies of torpedo injuries that took place at Fort Wagner. All five Union soldiers had been wounded by landmines after the fall of the fort. When Corp. William H. Rich of the Fourth New Hampshire was admitted to the field hospital at noon on September 8, 1863, "the tarsal bones were extensively crushed, and the soft parts were in such a condition that no operation could be performed at the ankle-joint." The result was amputation of the lower third of his left leg. The surgeon used five ligatures to close blood vessels; they came away on October 5, "and the patient was entirely well with a good stump, on the 10th of November."[6]

Pvt. George Wagner of the New York Independent Battalion of Infantry (Les Infants Perdue) was admitted at 7 a.m. on September 9 with a "compound comminuted fracture of both bones of right leg in their lower third. Above the point of fracture a large flap of integument on the calf of the leg had been made, and it was blackened by powder." Assistant surgeon W. H. Finn cut the leg in the middle, tying off six blood vessels. By day six of Wagner's hospitalization, "some sloughing of the integuments of the calf, with a discharge of offensive pus [occurred], but in a few days healthy granulations sprung up under the use of creosote dressing." The last ligature "came away" on the twenty-third day of his hospitalization. Wagner was transferred north on October 10, 1863, with his "stump nearly healed." Twenty-seven years old when he had enlisted in January 1862, Wagner was discharged from the army on June 23, 1864.[7]

Sgt. Thomas Mack of the Fourth New Hampshire was admitted at noon on September 9. "The right foot and lower third of the leg were swollen, livid, and cold; a fracture of the fibula, about four inches above the ankle-joint, and one of the tibia lower down, could be detected. The integument was unbroken, and he suffered severe pain." Sixty hours later, the surgeon made an incision to relieve clotted blood, and forty-eight hours after that, the limb had warmed up enough to inspire hope of saving it. But "mortification set in" on the night of September 13 and, by the next day, had begun to spread up from the ankle. The surgeon amputated the limb, using eight ligatures, one of them on the femoral vein. Mack "was much exhausted by the operation and by his previous suffering." Morphine helped ease the pain, but he died on the morning of September 16.[8]

Pvt. Peter Riley of Company M, Third Rhode Island Heavy Artillery was admitted at 1 p.m. on September 19 with a "very extensive torpedo wound of the right lower limb. The foot, with the exception of a small fragment of calcaneum [the large bone of the heel], had been blown away; the missile—supposed to be part of the metallic exploding apparatus—had passed up the leg, leaving only the soleus [a large muscle on the back of the leg] and gastrocnemius muscles [another muscle on the back of the leg] and the integument [skin], and exposing the posterior tibial vessels [blood vessels on the lower back of the leg] and nerve. The entire anterior [front] parts of the leg with the bones were carried away, a small portion of the head of the tibia [the main bone between the knee and the ankle] alone remaining. The knee-joint was extensively fractured, and the internal semi-lunar cartilage [the cartilage in the middle of the knee joint] had disappeared. The thigh-bone was fractured obliquely three inches above the knee, and bared to the extent of about six inches more, the periosteum [membranous tissue covering the surface of bones] being much bruised." Surgeons cut off the limb at the thigh using six ligatures, including one for the femoral vein. But that vein continued to bleed. On the third day of his hospitalization, Riley suffered "slight traumatic delirium." He was sent to Beaufort, South Carolina, on September 30, where the ligatures came away on October 9. Within a month Riley had "entirely recovered" and was later discharged from the army.[9]

Pvt. John R. Fordice of the Thirty-ninth Illinois was admitted on September 12 with "comminuted fractures of both bones of the right leg extending into the knee-joint, with great destruction of the soft parts." The surgeon amputated the leg at the thigh using seven ligatures. By the twenty-first day of hospitalization, the ligature on the femoral artery came away, and Fordice recovered by the end of his seventh week in medical care. He was discharged from the army on June 20, 1864.[10]

These detailed case studies of Union torpedo casualties are rare, but they are supplemented by one Confederate case study. Assistant surgeon Jack Bryant Stinson of the Forty-first Alabama provided details about one of his men who had been injured by a Rebel torpedo during the Petersburg campaign. He described this case in an article published in the *Southern Practitioner* in 1900. For some reason, however, the journal spelled his name Stimson. The surgeon reported that Sgt. Jerry Burkhalter of Company D, Forty-first Alabama decided to take a shortcut while going out to or returning from the picket line and walked along the Norfolk and Petersburg Railroad, where the track ran through the Confederate line. Burkhalter stepped on a landmine placed between the ties in an incident already referred to in chapter 6.[11]

Stinson described the effect of the mine on Burkhalter. "The shoe was a little powder burned on the bottom and a split about an inch long, as if cut with a knife, was on the front. This was the only injury to the shoe, and the skin was not broken on the foot, but every bone in the foot, up to several inches above the ankle, was crushed by the concussion." Despite this terrible injury, Burkhalter stubbornly refused to let Stinson amputate and died about ten days later.[12]

The cases reported by Gross and Stinson include six named men who were victims of landmines in the Civil War. In addition to those half a dozen names, other accounts reveal the names of sixteen other landmine victims of the conflict. The total of 25 named casualties of torpedoes that were planted on the battlefield is only a small fraction (19.5 percent) of the 128 estimated casualties of battlefield landmines.

Notes

PREFACE

1. Croll, *History of Landmines*, xi.
2. Croll, *History of Landmines*, ix, 1–8; Schneck, "Origins of Military Mines: Part 1," 49–51; Youngblood, *Development of Mine Warfare*, 1.
3. *OR*, vol. 28, pt. 2, 37; Delafield, *Report on the Art of War*, 109.
4. "Torpedo," https://en.wikipedia.org/wiki/torpedo; Croll, *History of Landmines*, ix.
5. Barnes, *Submarine Warfare*, 16n.
6. *Oxford English Dictionary*, vol. 18, 270.
7. *Gazette of the United States*, October 7, 1789.
8. Prescott, *History, Theory and Practice*, 35–37.
9. Charles W. Phillips letter, May 6, 1862, quoted in *History of the Fifth Massachusetts Battery*, 248; Howe, ed., *Marching with Sherman*, 161; Copp, *Reminiscences of the War*, 257; Crooke, *Twenty-First Regiment*, 146.
10. King, *Torpedoes: Their Invention and Use*, 85.
11. *Oxford English Dictionary*, vol. 18, 270.

CHAPTER 1

1. Schneck, "Origins of Military Mines: Part 1," 52; Youngblood, *Development of Mine Warfare*, 5–6.
2. Jones, *Malice Aforethought*, 11–12; Youngblood, *Development of Mine Warfare*, 6; Schneck, "Origins of Military Mines: Part 1," 52.
3. Schneck, "Origins of Military Mines: Part 1," 52; Pleydell, *Essay on Field Fortification*, 167–70, 172.
4. Pleydell, *Essay on Field Fortification*, 175–76, 178–79, 183.
5. Tielke, *Field Engineer*, vol. 2, 36–37.
6. Croll, *History of Landmines*, 9; Barnes, *Submarine Warfare*, 16; Schneck, "Origins of Military Mines: Part 1," 52.
7. Smith, *War with Mexico*, vol. 2, 151; Ramsey, trans., *Other Side*, 355.

8. Smith, *War with Mexico*, vol. 2, 155–57; Williams, ed., *With Beauregard in Mexico*, 81.

9. Smith, *War with Mexico*, vol. 2, 410n; Ramsey, trans., *Other Side*, 363.

10. Schneck, "Origins of Military Mines: Part 1," 51; Croll, *History of Landmines*, 9.

11. Halleck, *Elements of Military Art and Science*, 360, 363; Scott, *Military Dictionary*, 317–18.

12. Waters, "'Deception Is the Art of War,'" 31; Heitman, *Historical Register*, vol. 1, 813; William E. Bergin to J. Porter Donnell, October 11, 1951, Gabriel J. Rains Service Record, M331, NARA; Crist, ed., *Papers of Jefferson Davis*, vol. 1, 357, 381.

13. Rains to David E. Twiggs, May 29, 1840, Gabriel J. Rains Service Record, RG 94, NARA.

14. Rains to David E. Twiggs, May 29, 1840, Gabriel J. Rains Service Record, RG 94, NARA; Rains, "Torpedoes," 257–60; "Invention of Torpedoes," 734; Mahon, *History of the Second Seminole War*, 375; Waters, "'Deception Is the Art of War,'" 32; Rains to Joel R. Poinsett, July 30, 1840: Poinsett to Rains, August 20, 1840, Gabriel J. Rains Papers, SCHS.

15. Schiller, ed., *Confederate Torpedoes*, 3; Youngblood, *Development of Mine Warfare*, 25.

16. Schiffer, *Power Struggles*, 13–14; "Voltaic Pile," https://en.wikipedia.org/wiki/Voltaic_pile; Lockwood, *Electricity, Magnetism, and Electric Telegraphy*, 24–26.

17. Kochan and Wideman, *Civil War Torpedoes*, 335, 337, 567–68, 582; Schiffer, *Power Struggles*, 14, 19, 75–79; "William Robert Grove," https://en.wikipedia.org/wiki/William_Robert_Grove; Lockwood, *Electricity, Magnetism, and Electric Telegraphy*, 28.

18. Schiffer, *Power Struggles*, 22, 27–28, 32, 34, 42–53, 49, 51–52; Schellen, *Magneto-Electric and Dynamo-Electric Machines*, 15–16; Lockwood, *Electricity, Magnetism, and Electric Telegraphy*, 52–53, 59–62.

19. Hutcheon, *Robert Fulton*, 106, 109, 149; King, *Torpedoes*, 85; Porter, "Torpedo Warfare," 215–18, 220; Lundeberg, *Samuel Colt's Submarine Battery*, iii, 25–26, 37, 42–46.

20. Lundeberg, *Samuel Colt's Submarine Battery*, 4–5; Schiffer, *Power Struggles*, 120–21; King, *Torpedoes*, 18, 77; "Charles Wheatstone," https://en.wikipedia.org/wiki/Charles_Wheatstone; Lockwood, *Electricity, Magnetism, and Electric Telegraphy*, 61.

21. Delafield, *Report on the Art of War*, 36, 109; "Immanuel Nobel the Younger," https://en.wikipedia.org/wiki/Immanuel_Nobel; Youngblood, *Development of Mine Warfare*, 28, 31; "Moritz Hermann von Jacobi," https://en.wikipedia.org/wiki/Moritz_von_Jacobi.

22. Youngblood, *Development of Mine Warfare*, 28–29.

23. Baumgart, *Crimean War*, 150–51.

24. Fletcher and Ishchenko, *Crimean War*, 391; Robins, ed., *Captain Dunscombe's Diary*, 121–22; Pack, *Sebastopol Trenches*, 146, 149, 151, 153.

25. Hibbert to mother, June 14, 1855, Letters and Papers of Colonel Hugh Robert Hibbert, CALS.

26. Pack, *Sebastopol Trenches*, 154–57; Hibbert to mother, June 14, 1855, Letters and Papers of Colonel Hugh Robert Hibbert, CALS.

27. Delafield, *Report on the Art of War*, 109–10; Bentley, ed., *Russell's Dispatches*, 222; Pack, *Sebastopol Trenches*, 154.

28. Bentley, ed., *Russell's Dispatches*, 221–22; Pack, *Sebastopol Trenches*, 168; Delafield, *Report on the Art of War*, 109.

29. Delafield, *Report on the Art of War*, 110–13.

30. Youngblood, *Development of Mine Warfare*, 29–30.

31. Youngblood, *Development of Mine Warfare*, 30, 33.

32. Delafield, *Report on the Art of War*, 112.

33. Porter, "Torpedo Warfare," 221.

34. *New York Herald*, June 3, 1850.

35. *Daily Comet* (Baton Rouge, Louisiana), November 20, 1852.

36. *National Republican* (Washington, DC), July 6, 1861; King, *Torpedoes*, 85.

CHAPTER 2

1. Cutrer, *Theater of a Separate War*, 54–56.

2. Mulligan, "Siege of Lexington," 307n, 308–09; Crowell, "With Mulligan at Lexington," 45.

3. Cutrer, *Theater of a Separate War*, 54–56.

4. Ammen, "DuPont and the Port Royal Expedition," 678.

5. Denison, *Shot and Shell*, 49–50.

6. Ammen, "DuPont and the Port Royal Expedition," 686–87.

7. Ammen, "DuPont and the Port Royal Expedition," 687; *OR*, vol. 6, 18–20.

8. Hess, *Civil War in the West*, 11–17.

9. Nichols, *Confederate Engineers*, 51; *ORN*, vol. 22, 806; *OR*, vol. 52, pt. 2, 230; Requisition on Ordnance Department, July 3, 1862, Marshall McDonald Service Record, NARA.

10. Polk to wife, January 6, 1862, Leonidas Polk Papers, US; Polk, *Leonidas Polk*, vol. 2, 65; Simon, ed., *Papers of Ulysses S. Grant*, vol. 3, 375.

11. *OR*, vol. 7, 438.

12. *OR*, vol. 7, 682–83.

13. *OR*, vol. 7, 437.

14. *OR*, vol. 7, 436; Henri Lovie sketch and letter, March 8, 1862, in *Frank Leslie's Illustrated Newspaper*, March 29, 1862, 308, 317–18.

15. Henry Lovie letter, March 8, 1862, in *Frank Leslie's Illustrated Newspaper*, March 29, 1862, 318; correspondence dated March 9, 1862, in *Chicago Times*, reprinted in *Wisconsin Daily Patriot*, March 12, 1862.

16. Correspondence dated March 9, 1862, in *Chicago Times*, reprinted in *Wisconsin Daily Patriot*, March 12, 1862.

17. "Infernal Machines in the Mississippi," 210–11.

18. Lossing, *Pictorial Field Book*, vol. 2, 237n; Kochan and Wideman, *Civil War Torpedoes*, 244.

19. *Official Military Atlas*, plate 5, no. 2.
20. Kerner, ed., "Diary of Edward W. Crippen," 236.
21. *Frank Leslie's Illustrated Newspaper*, March 29, 1862, 317.
22. Thomas B. Beggs Diary, December 2, 1862, TSLA.

CHAPTER 3

1. Waters, "'Deception Is the Art of War,'" 32; Waters and Brown, *Gabriel Rains*, 23, 25; "Invention of Torpedoes," 234.
2. Cards and Magruder to Randolph, April 8, 1862, Gabriel J. Rains Service Record, M331, NARA; *OR*, vol. 4, 668; *OR*, vol. 9, 37.
3. Sears, *To the Gates of Richmond*, 40–62.
4. *New York Times*, May 8, 1862.
5. Ratchford, "More of Gen. Rains," 283.
6. Nanzig, ed., *Civil War Memoirs*, 36; Bryan and Lankford, eds., *Eye of the Storm*, 60; Wittenberg, ed., *"We Have It Damn Hard,"* 29; Aldrich, *History of Battery A*, 75; *History of the Fifth Massachusetts Battery*, 245–46, 248; Donald, ed., *Gone for a Soldier*, 64; Griffin, ed., *Three Years a Soldier*, 44; Taylor, ed., *Saddle and Saber*, 29.
7. *OR*, vol. 11, pt. 1, 398, 400, 402; Parker, *Twenty-Second Massachusetts*, 94–95.
8. Parker, *Twenty-Second Massachusetts*, 95, 97; *OR*, vol. 11, pt. 1, 400.
9. Floyd, *History of the Fortieth*, 142, 386, 388; *OR*, vol. 11, pt. 1, 401; Charles E. Halsey to sister, May 4, 1862, Bell Halsey Miller Papers, DU.
10. *OR*, vol. 11, pt. 1, 511, 557; William E. Sleight Reminiscences, 2, BHL-UM; Harry A. Purviance to *Reporter*, May 13, 1862, in Dickey, *Eighty-fifth Regiment*, 36–37.
11. Banes, *History of the Philadelphia Brigade*, 56; Bombaugh, "Extracts from a Journal," 316–17; Buckman to Friend Smith, May 11, 1862, George R. Buckman Civil War Papers, MHS.
12. Peleg W. Blake letter, May 5, 1862, and David Henry Grows journal, May 5, 1862, and Charles A. Phillips letter, May 6, 1862, in *History of the Fifth Massachusetts Battery*, 244–45, 248; "Letter from a Sharpshooter," May 4, 1862, *Rutland Weekly Herald*, May 15, 1862; Marks, *Peninsula Campaign*, 150; Hays, comp., *Under the Red Patch*, 83; Bryan and Lankford, eds., *Eye of the Storm*, 59; Lord, *Civil War Collector's Encyclopedia*, 171.
13. Howard, "Yorktown Evacuated," scrapbook clipping, Box 6, Oliver O. Howard Papers, LMU; Bryan and Lankford, eds., *Eye of the Storm*, 59; *SOR*, pt. 1, vol. 2, 45; Buckman to Friend Smith, May 11, 1862, George R. Buckman Civil War Papers, MHS; Howard, *Autobiography*, vol. 1, 218. Other sources claim Lathrop was killed in the streets of Yorktown. See Parker, *Twenty-Second Massachusetts*, 95–96; *OR*, vol. 11, pt. 1, 400; Special correspondent letter, May 4, 1862, *New York Herald*, May 7, 1862; Plum, *Military Telegraph*, vol. 1, 145.
14. Reichardt, *Diary of Battery A*, 41; Moses Hill to wife, May 11, 1862, in Roche, ed., *"Our Aim Was Man,"* 155.
15. *SOR*, pt. 1, vol. 2, 45; Nevins, ed., *Diary of Battle*, 45.

16. Diary, May 8, 1862, James M. Uhler Papers, USAMHI; David Henry Grows journal, May 5, 1862, and Charles A. Phillips letter, May 6, 1862, in *History of the Fifth Massachusetts Battery*, 245, 248; W. D. O'Brien, "They Planted Shells," *National Tribune*, June 17, 1886.

17. Howard, "Yorktown Evacuated," scrapbook clipping, Box 6, Oliver O. Howard Papers, LMU; Howard, *Autobiography*, vol. 1, 222–23; Howard to wife, May 7, 1862, O. O. Howard Papers, BC.

18. Rhodes, *Battery B*, 84.

19. Bombaugh, "Extracts from a Journal," 316; Howard, "Yorktown Evacuated," scrapbook clipping, Box 6, Oliver O. Howard Papers, LMU.

20. Correspondent of *Cincinnati Commercial*, May 14, 1862, in *Daily Argus* (Rock Island, Illinois), May 17, 1862.

21. Donald, ed., *Gone for a Soldier*, 64.

22. Bryan and Lankford, eds., *Eye of the Storm*, 64.

23. Rains, "Torpedo Book," in Schiller, ed., *Confederate Torpedoes*, 59; Rains to Walthall, June 21, 1879, in Waters and Brown, *Gabriel Rains*, 129.

24. Rains, "Torpedo Book," in Schiller, ed., *Confederate Torpedoes*, 59; Rains to Walthall, June 21, 1879, in Waters and Brown, *Gabriel Rains*, 129.

25. Charles A. Phillips letter, May 6, 1862, in *History of the Fifth Massachusetts Battery*, 248; Bombaugh, "Extracts from a Journal," 318.

26. Malles, ed., *Bridge Building in Wartime*, 65; Bryan and Lankford, eds., *Eye of the Storm*, 60.

27. Special correspondent letter, May 4, 1862, *New York Herald*, May 7, 1862.

28. Special correspondent letter, May 4, 1862, *New York Herald*, May 7, 1862; Bombaugh, "Extracts from a Journal," 318; Floyd, *History of the Fortieth (Mozart) Regiment*, 142.

29. *OR*, vol. 11, pt. 1, 313; Bombaugh, "Extracts from a Journal," 317; Alexander, "Sketch of Longstreet's Division," 38; Youngblood, *Development of Mine Warfare*, 41.

30. Sears, ed., *Civil War Papers*, 254, 256.

31. Diary, May 8, 1862, James M. Uhler Papers, USAMHI; Roche, ed., *"Our Aim Was Man,"* 153; Buckman to Friend Smith, May 11, 1862, George R. Buckman Civil War Papers, MHS; Mehaffy to mother, May 11, 1862, Calvin D. Mehaffy Papers, WLC-UM; Messent and Courtney, eds., *Civil War Letters*, 124.

32. *New York Herald*, May 6, 1862; Harry A. Purviance to *Reporter*, May 13, 1862, in Dickey, *Eighty-fifth Regiment*, 37; St. Clairsville, Ohio, *Belmont Chronicle*, May 8, 1862; Ebensburg, Pennsylvania, *The Alleghanian*, May 22, 1862.

33. Special correspondent letter, May 4, 1862, *New York Herald*, May 7, 1862; Buckman to Friend Smith, May 11, 1862, George R. Buckman Civil War Papers, MHS; *Nashville Union*, May 31, 1862.

34. Byrnes to Florence Clark, May 15, 1862, William Byrnes Letter, MML.

35. Adams, *Trooper's Adventures*, 382; Marks, *Peninsular Campaign*, 151; Banes, *History of the Philadelphia Brigade*, 56; Floyd, *History of the Fortieth (Mozart) Regiment*, 142; Hyde, *Following the Greek Cross*, 48.

36. Alexander, "Sketch of Longstreet's Division," 38; Davis, *Campaign from Texas to Maryland*, 30; Lasswell, ed., *Rags and Hope*, 76; Nanzig, ed., *Civil War Memoirs*, 36.
37. *OR*, vol. 11, pt. 3, 509.
38. Sorrel, *Recollections of a Confederate*, 69.
39. *OR*, vol. 11, pt. 3, 510.
40. *OR*, vol. 11, pt. 3, 509–10.
41. *OR*, vol. 11, pt. 3, 509.
42. *OR*, vol. 11, pt. 3, 510; *OR*, vol. 11, pt. 1, 605.
43. *OR*, vol. 11, pt. 3, 511.
44. *OR*, vol. 11, pt. 3, 516.
45. *OR*, vol. 11, pt. 3, 516.
46. Rains, "Torpedo Book," in Schiller, ed., *Confederate Torpedoes*, 56, 59.
47. *OR*, vol. 11, pt. 3, 517.
48. Rains to brother, July 5, 1862, Gabriel J. Rains Papers, SCHS; *OR*, vol. 11, pt. 3, 510.
49. *OR*, vol. 11, pt. 3, 510; Rains to brother, July 5, 1862, Gabriel J. Rains Papers, SCHS.
50. *New York Times*, May 8, 1862; Tidwell, Hall, and Gaddy, *Come Retribution*, 161; *OR*, vol., 28, pt. 1, 523; Lewis, ed., "William Fisher Plane," 221.
51. Rains, "Torpedo Book," in Schiller, ed., *Confederate Torpedoes*, 59.
52. Minnich, "Incidents of the Peninsular Campaign," 53; *OR*, vol. 11, pt. 1, 970.
53. Ratchford, "More of Gen. Rains," 283.

CHAPTER 4

1. Sears, *To the Gates*, 126, 128; *OR*, vol. 11, pt. 1, 943–45, 969–70, 976.
2. *OR*, vol. 11, pt. 1, 944, 974, 976; Sears, *To the Gates*, 131–33.
3. *OR*, vol. 11, pt. 3, 605; Crist, ed., *Papers of Jefferson Davis*, vol. 8, 225–26.
4. Rains, "Torpedoes," 260; Waters and Brown, *Gabriel Rains*, 131.
5. *OR*, vol. 11, pt. 3, 608; Waters, "'Deception Is the Art of War,'" 39.
6. Rains to brother, July 5, 1862, Gabriel J. Rains Papers, SCHS.
7. "George Washington Rains," in Malone, ed., *Dictionary of American Biography*, Vol. 15, 328–29; Rains to brother, July 5, 1862, Gabriel J. Rains Papers, SCHS; Waters and Brown, *Gabriel Rains*, 20–21.
8. Waters, "'Deception Is the Art of War,'" 41; *OR*, vol. 42, pt. 3, 1219–20.
9. Crist, ed., *Papers of Jefferson Davis*, vol. 8, 515.
10. Waters, "'Deception Is the Art of War,'" 41–43; Wideman, *Sinking of the USS Cairo*, 60; Kochan and Wideman, *Civil War Torpedoes*, 57–58.
11. Miers, ed., *Rebel War Clerk's Diary*, vol. 1, 233–34, 245.
12. Miers, ed., *Rebel War Clerk's Diary*, vol. 1, 246.
13. *OR*, vol. 11, pt. 1, 349; Rains to brother, July 5, 1862, Gabriel J. Rains Papers, SCHS; Rains, "Torpedo Book," in Schiller, ed., *Confederate Torpedoes*, 59; Miers, ed., *Rebel War Clerk's Diary*, vol. 1, 246.

14. Rains, "Torpedo Book," in Schiller, ed., *Confederate Torpedoes*, 63; Lord, *Civil War Collector's Encyclopedia*, 173; Croll, *History of Landmines*, 17; Kochan and Wideman, *Civil War Torpedoes*, 109–11, 118; Jones, *Artillery Fuses*, 122, 124–25; photograph in Schiller, ed., *Confederate Torpedoes*, 188; Bell, *Civil War Heavy Explosive Ordnance*, 474–76.

15. *OR*, vol. 18, 1022, 1082–83; *OR*, vol. 52, pt. 2, 508.

16. *OR*, vol. 18, 1082; Waters, "'Deception Is the Art of War,'" 45.

17. Rowland, ed., *Jefferson Davis*, vol. 5, 504.

18. *OR*, vol. 24, pt. 3, 919; Crist, ed., *Papers of Jefferson Davis*, vol. 9, 189.

19. *OR*, vol. 18, 1082.

20. *OR*, vol. 18, 1082; *OR*, vol. 24, pt. 1, 220.

21. *OR*, vol. 18, 1082–83.

22. Rowland, ed., *Jefferson Davis*, vol. 5, 504.

23. Rowland, ed., *Jefferson Davis*, vol. 5, 504.

24. Vouchers, no dates, in Gabriel J. Rains Service Record, M331, NARA.

25. Rains to Vance, June 29, 1863, Gabriel J. Rains Letter, SIU.

26. Rains to Vance, June 29, 1863, Gabriel J. Rains Letter, SIU.

27. Robinson Morning Report, July 2, 1863, Samuel H. Lockett Papers, UNC; Hogane, "Reminiscences of the Siege of Vicksburg, Pt. 2," 295.

28. *SOR*, pt. 1, vol. 4, 749, 802; "Fortification and Siege of Port Hudson," 334; Millett, "At Port Hudson: A Boy of the 24th Me. Has Some Exciting Adventures."

29. Crist, ed., *Papers of Jefferson Davis*, vol. 9, 269; Rains, "Torpedo Book," in Schiller, ed., *Confederate Torpedoes*, 68.

30. VIDI (alias of H. F. Scaife) account, quoted in Stone, ed., *Wandering to Glory*, 118; Joyce, "Infantry Stampede," 223.

31. *OR*, vol. 24, pt. 2, 536; Curtis P. Lacey Diary, July 18, 1863, NL; John Merrilles Diary, July 17, 1863, CHM; Jones to parents, July 22, 1863, John G. Jones Civil War Letters, WHS; Alfred A. Rigby Diary, July 17, 1863, EU; diary, July 17, 1863, Hugh Boyd Ewing Papers, OHS.

32. *History of the Forty-Sixth Regiment Indiana*, 69; Rains, "Torpedo Book," in Schiller, ed., *Confederate Torpedoes*, 68.

33. "The Capture of Jackson," in Moore, *Rebellion Record*, vol. 7, 351.

34. Furney, ed., *Reminiscences of the War*, 127–28.

35. Curtis P. Lacey Diary, July 18, 1863, NL; diary, July 18, 1863, Spencer S. Kimbell Papers, CHM; John Merrilles Diary, July 17, 1863, CHM.

36. Henry S. Nourse, "From Young's Point to Atlanta," in *Story of the Fifty-Fifth Regiment*, 261; Alfred A. Rigby Diary, July 17, 1863, EU.

37. Rains, "Torpedo Book," in Schiller, ed., *Confederate Torpedoes*, 68–69.

38. *OR*, vol. 24, pt. 2, 537.

39. Miers, ed., *Rebel War Clerk's Diary*, vol. 2, 8; Welburn J. Andrews account, quoted in Stone, ed., *Wandering to Glory*, 118; Davis, *Rise and Fall of the Confederate Government*, vol. 2, 424–25.

CHAPTER 5

1. Hess, *Field Armies*, 251.
2. Hess, *Field Armies*, 251–54.
3. Hess, *Field Armies*, 254–58.
4. *OR*, vol. 28, pt. 1, 58, 92–93; *OR*, vol. 28, pt. 2, 186, 195, 226.
5. *OR*, vol. 28, pt. 1, 523.
6. *OR*, vol. 28, pt. 1, 523; *OR*, vol. 28, pt. 2, 213; Rains to W. T. Walthall, June 21, 1879, quoted in Waters and Brown, *Gabriel Rains*, 130.
7. Beauregard to General, July 25, 1863, G. T. Beauregard Papers, USC.
8. *OR*, vol. 28, pt. 1, 523.
9. *OR*, vol. 28, pt. 2, 297, 323–24, 332; *OR*, vol. 28, pt. 1, 523.
10. Trautmann, ed., *A Prussian Observes*, 97; Gordon, *War Diary*, 220.
11. Schneck, "Origins of Military Mines: Part II," 47.
12. *OR*, vol. 28, pt. 1, 24, 204.
13. *OR*, vol. 28, pt. 1, 24.
14. *OR*, vol. 28, pt. 1, 294–96; Price, *History of the Ninety-Seventh Regiment*, 191.
15. *OR*, vol. 28, pt. 1, 296; *Harper's Weekly*, September 19, 1863; Kochan and Wideman, *Civil War Torpedoes*, 140.
16. *OR*, vol. 28, pt. 1, 296, 396, 501.
17. Hyde, *History of the One Hundred and Twelfth Regiment*, 56; *OR*, vol. 28, pt. 1, 297; George Benson Fox to father and mother, September 9, 1863, Fox Collection, CinHM.
18. *OR*, vol. 28, pt. 1, 298.
19. *OR*, vol. 28, pt. 1, 301, 339–40.
20. *OR*, vol. 28, pt. 1, 26, 301.
21. *OR*, vol. 28, pt. 1, 310.
22. *OR*, vol. 28, pt. 1, 310, 312.
23. Gross, "On Torpedo Wounds," 370; Kochan and Wideman, *Civil War Torpedoes*, 137.
24. *OR*, vol. 28, pt. 1, 312.
25. *OR*, vol. 28, pt. 1, 312; Hyde, *History of the One Hundred and Twelfth Regiment*, 56.
26. *OR*, vol. 28, pt. 1, 446, 479, 512.
27. *OR*, vol. 28, pt. 1, 24, 37, 296; Gillmore, "Army before Charleston," 63; Davis, *History of the 104th Pennsylvania*, 263.
28. *OR*, vol. 28, pt. 1, 296: *OR*, vol. 28, pt. 1, 501.
29. *OR*, vol. 28, pt. 1, 469–70, 478; Wise, *Gate of Hell*, 135.
30. Hess, *Field Armies*, 280–82; George Benson Fox to Mary, September 15, 1863, Fox Collection, CinHM.
31. *OR*, vol. 28, pt. 2, 92; Gross, "On Torpedo Wounds," 371–72; Palladino, ed., *Diary of a Yankee Engineer*, 34–35. There is some evidence that the Confederates rigged booby traps in Fort Wagner, but it is not convincing. See Gordon, *War Diary*, 216, 220.
32. *OR*, vol. 28, pt. 2, 291.

33. *OR*, vol. 28, pt. 2, 324; Miers, ed., *Rebel War Clerk's Diary*, vol. 2, 31.
34. Palladino, ed., *Diary of a Yankee Engineer*, 34–35; Gordon, *War Diary*, 220.
35. Cutrer, *Theater of a Separate War*, 304–06, 309.
36. Ragan, *Confederate Saboteurs*, 1–2, 12–13, 21–22.
37. *OR*, vol. 34, pt. 2, 854–55; Perry, *Infernal Machines*, 46.
38. *OR*, vol. 26, pt. 1, 419, 446; Thomas J. Parker, "The Capture of Fort Esperanza, Texas," *National Tribune*, August 16, 1883; Marshall, *Army Life*, 218–25.
39. *OR*, vol. 26, pt. 1, 416–18.
40. Baker to Allen, October 2, 1865, quoted in King, *Torpedoes*, 6; *OR*, vol. 34, pt. 2, 854.
41. Baker to Allen, October 2, 1865, quoted in King, *Torpedoes*, 5–6.
42. Baker to Allen, October 2, 1865, quoted in King, *Torpedoes*, 6.
43. Wheeler to Malinda, December 3, 1863, George Wheeler Papers, NC.
44. Chace account, quoted in Cheney, *Fourteenth Regiment Rhode Island Heavy Artillery*, 20–22.
45. *OR*, vol. 26, pt. 2, 298–99.
46. Wheeler to Malinda, December 3, 1863, George Wheeler Papers, NC.

CHAPTER 6

1. *OR*, vol. 26, pt. 2, 136; Crist, ed., *Papers of Jefferson Davis*, vol. 9, 350.
2. *OR*, vol. 26, pt. 2, 180.
3. Elder to Mr. Gibbs, October 30, 1862, William D. Elder Papers, MDAH; Tidwell, Hall, and Gaddy, *Come Retribution*, 162–63.
4. Rowland, ed., *Jefferson Davis*, vol. 5, 596; "List of Commissioned Officers under command of Genl Rains," John Andrews to John M. Otey, January 13, 1864, Gabriel J. Rains Service Record, M331, NARA.
5. *OR*, vol. 32, pt. 2, 738–39; Waters, "'Deception Is the Art of War,'" 50.
6. Still, ed., "Civil War Letters of Robert Tarleton," 55–56.
7. A. L. Rives to J. E. Sullivan, April 29, 1864, Letters and Telegrams Sent by the Engineer Bureau of the Confederate War Department, NARA.
8. A. L. Rives to J. E. Sullivan, April 29, 1864, Letters and Telegrams Sent by the Engineer Bureau of the Confederate War Department, NARA.
9. *OR*, vol. 28, pt. 2, 371.
10. *OR*, vol. 28, pt. 2, 372.
11. Rhea, *The Battles for Spotsylvania Court House*, 189–212.
12. *OR*, vol. 36, pt. 1, 879; Wilson, "Cavalry of the Army of the Potomac," 50; Crowninshield, *History of the First Regiment of Massachusetts Cavalry*, 214; Hall, *History of the Sixth New York Cavalry*, 187; Hagemann, ed., *Fighting Rebels*, 234.
13. Sheridan, *Personal Memoirs*, vol. 1, 380; Foster, *Reminiscences and Record*, 74; Gracey, *Annals of the Sixth Pennsylvania Cavalry*, 244.
14. Hagemann, ed., *Fighting Rebels*, 235.

15. Charles E. Phelps, "Personal Recollections of the Wilderness Campaign," MaryHS; Gracey, *Annals of the Sixth Pennsylvania Cavalry*, 244; Griffin, ed., *Three Years a Soldier*, 223.

16. Wells A. Bushnell Memoirs, 269, WRHS.

17. Hagemann, ed., *Fighting Rebels*, 235.

18. Sheridan, *Personal Memoirs*, vol. 1, 380; "A Chat with Sheridan: The Story of His Raid Around Richmond, As Told by Himself," *National Tribune*, May 8, 1884; Nowlin, "Capture and Escape," 70; Pyne, *First New Jersey Cavalry*, 242; Gracey, *Annals of the Sixth Pennsylvania Cavalry*, 244.

19. Sheridan, *Personal Memoirs*, vol. 1, 380–81; Hall, *History of the Sixth New York Cavalry*, 187; Pyne, *First New Jersey Cavalry*, 242.

20. Trout, ed., *Memoirs of the Stuart Horse Artillery*, 88.

21. *OR*, vol. 36 pt. 2, 988; *OR*, vol. 36, pt. 3, 883.

22. *OR*, vol. 40, pt. 3, 764, 800.

23. *OR*, vol. 42, pt. 3, 1181–82.

24. Crist, ed., *Papers of Jefferson Davis*, vol. 11, 114.

25. *OR*, vol. 42, pt. 3, 1219–20.

26. Rains, "Torpedo Book," in Schiller, ed., *Confederate Torpedoes*, 60–62; Gallagher, ed., *Fighting for the Confederacy*, 443–44.

27. *OR*, vol. 42, pt. 3, 1181.

28. Hess, *In the Trenches*, 160–81.

29. Kochan and Wideman, *Civil War Torpedoes*, 236.

30. Hess, *In the Trenches*, 180; Jackson, *First Regiment Engineer Troops*, 107; Clifford Dickinson "Union and Confederate Engineering Operations at Chaffin's Bluff," 97–98, RNB; Shepley, "Incidents of the Capture of Richmond," 19; William Starr Basinger Reminiscences, 121, UGA; J. V. Bowles to sister, October 20, 1864, Bowles Family Papers, CWM.

31. Rains, "Torpedo Book," in Schiller, ed., *Confederate Torpedoes*, 57.

32. Rains, "Torpedo Book," in Schiller, ed., *Confederate Torpedoes*, 56–57.

33. Rains, "Torpedo Book," in Schiller, ed., *Confederate Torpedoes*, 57–58.

34. *OR*, vol. 42, pt. 3, 1219–20; Rains, "Torpedo Book," in Schiller, ed., *Confederate Torpedoes*, 55, 57.

35. *OR*, vol. 42, pt. 3, 645, 1181, 1219; Sommers, *Richmond Redeemed*, 109, 162, 165–66, 470, 584.

36. *OR*, vol. 42, pt. 3, 1220; Rains, "Torpedo Book," in Schiller, ed., *Confederate Torpedoes*, 17.

37. William Starr Basinger Reminiscences, 121, UGA; J. V. Bowles to sister, October 20, 1864, Bowles Family Papers, CWM; Carter Nelson Berkeley Minor "Record of the War," vol. 3, 58, 60, Minor Family Papers, UVA.

38. *OR*, vol. 42, pt. 3, 282, 613; *Private and Official Correspondence*, vol. 5, 354–55.

39. *OR*, vol. 42, pt. 3, 1220; Kochan and Wideman, *Civil War Torpedoes*, 261.

40. *SOR*, pt. 1, vol. 7, 282–83; Hess, *In the Trenches*, 44.

41. Stimson, "Three Unusual Gun-Shot Wounds," 364.

42. *OR*, vol. 46, pt. 2, 1,293; Clifford Dickinson "Union and Confederate Engineering Operations at Chaffin's Bluff," 98, RNB.

43. Eisenschiml, ed., *Vermont General*, 269.

44. *OR*, vol. 46 pt. 3, 212.

45. Abbot, *Siege Artillery in the Campaigns Against Richmond*, 131; Michie to Delafield, October 1865, in King, *Torpedoes*, 14–15.

CHAPTER 7

1. Rowland, ed., *Jefferson Davis*, vol. 8, 415–16.

2. *OR*, vol. 44, 867; Rowland, ed., *Jefferson Davis*, vol. 6, 410.

3. *OR*, vol. 44, 865, 866; Crist., ed., *Papers of Jefferson Davis*, vol. 11, 171; Beauregard to Cobb, November 18, 1864, Telegraph Book, October 18, 1864–March 16, 1865, Folder 6, Box 1, George W. Brent Papers, TU.

4. *OR*, vol. 44, 880, 885.

5. *OR*, vol. 44, 934; Beauregard to Hardee, December 3, 1864, Telegrams from December 3, 1864, to March 18, 1865, G. T. Beauregard Papers, LC.

6. *OR*, vol. 44, 71; Kirwan, ed., *Johnny Green*, 179.

7. *OR*, vol. 44, 410; Kirwan, ed., *Johnny Green*, 179.

8. *SOR*, pt. 2, vol. 1, 5–7; Hoole, *Alabama Tories*, 22, 40; Todd, *First Alabama Cavalry*, 10; *OR*, vol. 52, pt. 1, 649.

9. Cards and certified copy return of marriage, March 30, 1900, Francis W. Tupper Service Record, First Alabama Cavalry, NARA; Hoole, *Alabama Tories*, 127; Gordon Hickenlooper, ed. "The Reminiscences of General Andrew Hickenlooper, 1861–1865," 80, USAMHI.

10. Howe, ed., *Marching with Sherman*, 160; Gordon Hickenlooper, ed. "The Reminiscences of General Andrew Hickenlooper, 1861–1865," 80, USAMHI, 80.

11. Gordon Hickenlooper, ed. "The Reminiscences of General Andrew Hickenlooper, 1861–1865," 80, USAMHI, 81; John C. Van Duzer Diary, December 9, 1864, DU; Todd, *First Alabama Cavalry*, 10.

12. Gordon Hickenlooper, ed. "The Reminiscences of General Andrew Hickenlooper, 1861–1865," 80, USAMHI, 81.

13. John C. Van Duzer Diary, December 9, 1864, DU; Tupper to parents, December 24, 1864, Francis W. Tupper Papers, ALPL; Howe, ed., *Marching with Sherman*, 162.

14. Gordon Hickenlooper, ed. "The Reminiscences of General Andrew Hickenlooper, 1861–1865," 80–81, USAMHI.

15. Howe, ed., *Marching with Sherman*, 161; Sherman, *Memoirs*, vol. 2, 194.

16. Howe, ed., *Marching with Sherman*, 161–62; Sherman, *Memoirs*, vol. 2, 194; W. W. Lomax, "The Torpedoes at Savannah," *National Tribune*, December 26, 1901. There is some evidence that a local civilian informed Sherman he could identify the Confederate soldiers by name who had planted the torpedoes at Pooler Station and that those men were among the Seventeenth Corps pool of prisoners, but this seems a bit hard to believe. See L. B. March, "Hunting Torpedoes Near Savannah," *National Tribune*, August 14, 1902.

17. Clark, ed., *Downing's Civil War Diary*, 236; Throne, ed., "History of Company D," 77; Nichols, *Story of the Great March*, 86; *OR*, vol. 44, 34; Jackson, *Colonel's Diary*, 171.

18. Gould and Kennedy, eds., *Memoirs of a Dutch Mudsill*, 311.

19. Howe, ed., *Marching with Sherman*, 162.

20. *OR*, vol. 44, 71, 149; Kirwan, ed., *Johnny Green*, 179; *SOR*, pt. 2, vol. 76, 449, 469; Sharland, *Knapsack Notes*, 48.

21. Howe, ed., *Marching with Sherman*, 165.

22. Hedley, *Marching Through Georgia*, 322; Kirwan, ed., *Johnny Green*, 179.

23. Gordon Hickenlooper, ed. "The Reminiscences of General Andrew Hickenlooper, 1861–1865," 80, USAMHI, 81; Howe, ed., *Marching with Sherman*, 163–64.

24. Howard, "Incidents and Operations," 443; Howe, ed., *Marching with Sherman*, 161–62; Nichols, *Story of the Great March*, 86; Jackson, *Colonel's Diary*, 171.

25. Hess, *Civil War in the West*, 262–63; *OR*, vol. 44, 110.

26. George W. Anderson account, quoted in Jones, *Historical Sketch of the Chatham Artillery*, 141–42; and in Jones, *Siege of Savannah*, 126. In his memoirs, Tomb never admitted to placing landmines anywhere. He did not mention that he approached Jones with the idea of placing torpedoes in Sherman's path. "I reported to Flag Officer Tucker, and was sent on to Savannah to report to Major General Hardee for special service outside of that city," is all he wrote about the Pooler Station and Fort McAllister incidents. Even long after the war was over, Tomb was careful not to implicate himself in a deed he obviously knew was morally questionable. Campbell, ed., *Engineer in Gray*, 96.

27. *OR*, vol. 44, 61; William E. Strong, "Particulars of the Death of Maj. General James B. McPherson July 22, 1864, and An Account of the Capture of Fort McAllister December 13, 1864, for General Wm. T. Sherman Commanding, Army of the United States," William T. Sherman Papers, LC; Hazen, *Narrative of Military Service*, 333. For other sources that contain references to the placement of these torpedoes at Fort McAllister, see Howe, ed., *Marching with Sherman*, 189; William Ludlow to Delafield, September 1, 1865, in King, *Torpedoes*, 3; Nichols, *Story of the Great March*, 91; Diary, December 13, 1864, Wayne Johnson Jacobs Diaries and Lists, LSU; Livingston, *"Among the Best Men,"* 47, 104.

28. George W. Anderson account, quoted in Jones, *Historical Sketch of the Chatham Artillery*, 140–41; Strong, "Capture of Fort McAllister," 418; Howard, *Autobiography*, vol. 2, 88; Louis E. Lambert speech, quoted in *Ninth Reunion of the 37th Regiment*, 57. Gary Livingston identified Confederate Private Thomas J. Mills as the one who told Bremfoerder about the torpedoes planted in the causeway, but offered no indication of where he obtained that information. See Livingston, *"Among the Best Men,"* 91.

29. Hazen, *Narrative of Military Service*, 331–33; *OR*, vol. 44, 110; Castel, *Tom Taylor's Civil War*, 203–04.

30. *OR*, vol. 44, 110; Strong, "Capture of Fort McAllister," 413; Howard, *Autobiography*, vol. 2, 89; Magaw to sister, December 22, 1864, Theophilus M. Magaw Papers, HL.

31. "From the Diary of a Private," *New York Times*, June 11, 1893.

32. Diary, December 13, 1864, Wayne Johnson Jacobs Diaries and Lists, LSU; Lyman Hardman, quoted in Livingston, *"Among the Best Men,"* 104–05; L. C. Huffine, "Torpedoes at Fort McAllister," *National Tribune*, March 14, 1907.

33. *OR*, vol. 44, 110; diary, December 13, 1864, Wayne Johnson Jacobs Diaries and Lists, LSU; Henry Schmidt to wife, December 27, 1864, Schmidt Family Papers, FHS.

34. *OR*, vol. 44, 79, 111, 122; Hazen, *Narrative of Military Service*, 333; Lucas, *New History of the 99th Indiana*, 141; Sherman, *Memoirs*, vol. 2, 199; Connelly, *History of the Seventieth Ohio*, 138; John G. Brown, "From Atlanta to Chicago," in *Story of the Fifty-Fifth Regiment*, 398; diary, December 13, 1864, Wayne Johnson Jacobs Diaries and Lists, LSU; "Military History of Captain Thomas Sewell, Co. G, 127th Ill. Vol. Inf. During the War of the Rebellion, 1861 to 1865," Thomas Sewell Papers, DU; John C. Van Duzer Diary, December 13, 1864, DU.

35. "List of Severe wounds in 2nd Div. 15th A.C. rec'd at Statesboro Ga. Dec. 4th & in assault on Fort McAllister, Dec. 13, 1864, 'Sherman's March to the Sea,'" Series 1, Folder 457, Roll 3, United States Sanitary Commission Records, NYPL.

36. Sherman, *Memoirs*, vol. 2, 199; Howard, *Autobiography*, vol. 2, 91; Howard, "Incidents and Operations," 447; Harwell and Racine, eds., *Fiery Trail*, 72.

37. George W. Anderson account, quoted in Jones, *Historical Sketch of the Chatham Artillery*, 142–43; Livingston, *"Among the Best Men,"* 123; *ORN*, vol. 16, 362; diary, December 14, 1864, Edward E. Schweitzer Papers, HL; Diary, December 14, 1864, Wayne Johnson Jacobs Diaries and Lists, LSU; Magaw to sister, December 22, 1864, Theophilus M. Magaw Papers, HL; Saunier, ed., *History of the Forty-Seventh Regiment Ohio*, 373. One man insisted that Col. John M. Oliver, commander of the Third Brigade in Hazen's division, ordered the prisoners to dig up the mines. See William L. Johnson, "The Assault of Fort McAllister," *National Tribune*, July 19, 1883.

38. George W. Anderson account, quoted in Livingston, *"Among the Best Men,"* 116; "From the Diary of a Private," *New York Times*, June 11, 1893.

39. Howe, ed., *Marching with Sherman*, 190.

40. William Ludlow to Delafield, September 1, 1865, in King, *Torpedoes*, 3.

41. *OR*, vol. 44, 208, 831.

42. *OR*, vol. 44, 166; Walton, ed., *Behind the Guns*, 130.

43. William Ludlow to Delafield, September 1, 1865, in King, *Torpedoes*, 3; *OR*, vol. 44, 61.

44. Gordon Hickenlooper, ed. "The Reminiscences of General Andrew Hickenlooper, 1861–1865," 80, USAMHI, 81; Tupper to parents, December 24, 1864, Francis W. Tupper Papers, ALPL.

45. Tupper to parents, December 24, 1864, Francis W. Tupper Papers, ALPL.

46. Tupper to father, January 4, 15, 1865, Francis W. Tupper Papers, ALPL.

47. Tupper to father, January 4, 15, 1865, Francis W. Tupper Papers, ALPL.

48. Tupper to father, January 15, 1865, Francis W. Tupper Papers, ALPL.

49. Tupper to father, January 15, 23, 29, 1865, Francis W. Tupper Papers, ALPL.

50. Tupper to father, February 13, 1865, Francis W. Tupper Papers, ALPL.

51. Gordon Hickenlooper, ed. "The Reminiscences of General Andrew Hickenlooper, 1861–1865," 80, USAMHI, 81.

CHAPTER 8

1. Fonvielle, *Wilmington Campaign*, 36, 41; *OR*, vol. 46, pt. 1, 407.
2. Lamb, "Fort Fisher," 263, 266–67.
3. Fonvielle, *Wilmington Campaign*, 108–09, 133, 145, 178; Lamb, "Fort Fisher," 271–77.
4. Lamb, "Fort Fisher," 271–77.
5. Fonvielle, *Wilmington Campaign*, 197, 204, 211, 226.
6. Fonvielle, *Wilmington Campaign*, 234; *OR*, vol. 46, pt. 1, 441; Lamb, "Thirty-Sixth Regiment," 639–40.
7. Fonvielle, *Wilmington Campaign*, 247, 249, 252–58, 265; Lamb, "Thirty-Sixth Regiment," 643.
8. *ORN*, vol. 11, 591–92; A. L. Knowlton to Comstock, January 21, 1865, C. B. Comstock Papers, LC.
9. A. L. Knowlton to Comstock, January 21, 1865, C. B. Comstock Papers, LC; *OR*, vol. 46, pt. 2, 215–16; Fonvielle, *Wilmington Campaign*, 253.
10. *OR*, vol. 46, pt. 2, 215–16; *OR*, vol. 46, pt. 1, 408; Sumner, ed., *Diary of Cyrus B. Comstock*, 307; "Memoir on the Capture of Ft. Fisher, [January] 22, 65," C. B. Comstock Papers, LC; Fonvielle, *Wilmington Campaign*, 237.
11. *ORN*, vol. 11, 592.
12. *OR*, vol. 46, pt. 2, 216; Jones, *Artillery Fuses*, 144.
13. *OR*, vol. 46, pt. 2, 215–17; Kochan and Wideman, *Civil War Torpedoes*, 249–50; "Frederick Augustus Abel," https://en.wikipedia.org/wiki/Frederick_Abel.
14. Lamb, "Thirty-Sixth Regiment," 638; *ORN*, vol. 11, 592.
15. Fonvielle, *Wilmington Campaign*, 303; *OR*, vol. 46, pt. 1, 425, 427.
16. *OR*, vol. 46, pt. 1, 428.
17. *OR*, vol. 46, pt. 1, 429.
18. *OR*, vol. 46, pt. 1, 430–31; Fonvielle, *Wilmington Campaign*, 306.
19. *OR*, vol. 46, pt. 2, 1158–59.
20. *OR*, vol. 47, pt. 1, 419–20.
21. *OR*, vol. 47, pt. 1, 419–20, 426.
22. Hight manuscript, quoted in Stormont, comp., *Fifty-Eighth Regiment*, 467–68.
23. Ludlow to Delafield, September 1, 1865, quoted in King, *Torpedoes*, 3; "Operations in the Interior: Additional Particulars of the Movement," *New York Times*, February 17, 1865; Schneck, "Origins of Military Mines: Part I," 53; Hight manuscript, quoted in Stormont, comp., *Fifty-Eighth Regiment*, 467–68; *OR*, vol. 47, pt. 2, 202; *OR*, vol. 47, pt. 1, 1006.
24. Tourgée, *Story of a Thousand*, 357–58; *OR*, vol. 47, pt. 1, 429, 682, 761.
25. Smith, *History of the Seventh Iowa*, 213–14; Hight manuscript, quoted in Stormont, comp., *Fifty-Eighth Regiment*, 468.
26. *OR*, vol. 47, pt. 2, 1077; Charles Caley to Juliaette Carpenter Caley, February 2, 1865, Caley Family Correspondence, UND; *OR*, vol. 47, pt. 1, 420; Slocum, *Life and Services*, 256–57; Patrick and Willey, eds., *Fighting for Liberty and Right*, 305.
27. Smith, *History of the Seventh Iowa*, 213–14; *OR*, vol. 47, pt. 1, 429.
28. *OR*, vol. 47, pt. 2, 76.

29. Ellison, ed., *On to Atlanta*, 105.
30. Jones, *Artillery Fuses*, 129.
31. *OR*, vol. 47, pt. 2, 1, 299–1,300.
32. *OR*, vol. 47, pt. 2, 1,300.
33. *OR*, vol. 47, pt. 2, 1,239, 1,336.
34. *OR*, vol. 47, pt. 1, 911–13; Hall and Hall, *Cayuga in the Field*, 279–80.
35. *OR*, vol. 47, pt. 3, 704.
36. Rains to daughter, March 8, 1865, Gabriel J. Rains Papers, SCHS.
37. Harrison, "Capture of Jefferson Davis," 131–32.
38. Harrison, "Capture of Jefferson Davis," 132.
39. Holmes to Stringfellow, April 7, 1865, quoted in Waters and Brown, *Gabriel Rains*, 89.

CHAPTER 9

1. *OR*, vol. 39, pt. 2, 816; *OR*, vol. 45, pt. 1, 1230.
2. Bergeron, *Confederate Mobile*, 76, 79, 87, 120, 122, 124–25.
3. Hearn, *Mobile Bay*, 81–136.
4. O'Brien, *Mobile, 1865*, 34, 37, 39, 47, 57, 64; Hearn, *Mobile Bay*, 146–47, 150–51, 154–57, 162.
5. Wood, *A History of the Ninety-Fifth Regiment*, 165–67, 169.
6. Holbrook, *Narrative*, 188–89; Allen to Delafield, October 3, 1865, quoted in King, *Torpedoes*, 4; copy, unidentified diary, March 27, 1865, 130th Illinois Folder, VNMP.
7. Scott, comp., *Thirty-Second Iowa*, 334–35; diary, April 2, 1865, and Miller to wife, April 8, 1865, Monroe Joshua Miller Papers, MHM; Henry Fike to Cimbaline, April 4, 1865, Henry and Lucy Fike Papers, UK.
8. Moses A. Cleveland Journal, March 26, 1865, MHS; "Diary of John S. Morgan," 580–81; "Civil War Diary of William M. Macy," 193; George Carrington Diary, March 28, 1865, CMH; Evans to father, March 29, 1865, Thomas Evans Letters and Diary, EWU; Dame to Belle, April 4, 1865, Lorin Low Dame Correspondence and Journals, MHS.
9. Wilson, "The Campaign Ending with the Capture of Mobile," 18.
10. Stephenson, "Defence of Spanish Fort," 121; Hughes, ed., *Civil War Memoir*, 360; Little and Maxwell, *History of Lumsden's Battery*, 66; *OR*, vol. 49, pt. 2, 1181; *St. Louis Daily Missouri Democrat*, April 10, 1865.
11. Wilson, "The Campaign Ending with the Capture of Mobile," 18; Allen to Delafield, October 3, 1865, quoted in King, *Torpedoes*, 4; extract from Surg. C. B. White report, in *Medical and Surgical History*, vol. 2, 339; Andrews, *History of the Campaign of Mobile*, 160; Howard, *History of the 124th Regiment Illinois*, 296; Way, comp., *History of the Thirty-Third Regiment Illinois*, 228; Johnson, *Muskets and Medicine*, 219; Woods, *Services of the Ninety-Sixth Ohio*, 116–17.
12. Extract from Surg. C. B. White report, in *Medical and Surgical History*, vol. 2, 339; translation of Carl W. Bernhardt diary, April 5, 1865, Bernhardt-Campbell

Family Papers, USAMHI; "Diary of John S. Morgan," 585; Drish to wife, March 29, 1865, James F. Drish Papers, ALPL.

13. Hearn, *Mobile Bay*, 176; [Gibson] to Liddell, [March 27, 1865], and [Gibson] to Maury, April 1, 1865, and [Gibson] to Garner, April 3, 1865, and [Gibson] to Maury and Liddell, April 7, 1865, Register and Morning Report Book, Box 12, Folder 2, Civil War Papers, Battalion Washington Artillery Collection, TU.

14. *SOR*, pt. 1, vol. 7, 639–46, 955–63, 966. Liddell did not mention torpedoes in his memoirs when describing the Mobile campaign. See Hughes, ed., *Liddell's Record*, 187–97.

15. Allen to Delafield, October 3, 1865, quoted in King, *Torpedoes*, 4–5; Way, comp., *History of the Thirty-Third Regiment Illinois*, 228; Sperry, *History of the 33rd Iowa*, 154–55.

16. Holbrook, *Narrative*, 189; Winschel, ed., *Civil War Diary of a Common Soldier*, 149; Palfrey, "Capture of Mobile," 544; clipping from *Evening Gazette* (Burlington, Iowa), November 26, 1909, in A. V. Kendrick Reminiscences, MHM; W. S. Pierce, "In Command of a Springfield," *National Tribune*, August 25, 1910; Phil M. Wagner, "When At Spanish Fort: Amusing Reminiscences of the Time Just Before Evacuation," *National Tribune*, February 24, 1898.

17. Hearn, *Mobile Bay*, 187–88, 190–91; "Diary of John S. Morgan," 587.

18. "Diary of John S. Morgan," 587; Henry Fike to Cimbaline, April 10, 1865, Henry and Lucy Fike Papers, UK.

19. Sperry, *History of the 33rd Iowa*, 155; correspondence to *Cincinnati Gazette*, April 15, 1865, in *St. Louis Daily Missouri Democrat*, April 25, 1865.

20. "Diary of John S. Morgan," 587; Wilson, "The Campaign Ending with the Capture of Mobile," 17–18; Sperry, *History of the 33rd Iowa*, 155.

21. Yeary, comp., *Reminiscences*, 555; George W. Read, "Gen. Osterhaus Was Cool," *National Tribune*, September 15, 1904.

22. Hearn, *Mobile Bay*, 179, 181, 183; Moses A. Cleveland Journal, April 3, 1865, MHS; Trusty, "Private Smith Takes Mobile," 81; George Carrington Diary, April 3, 1865, CHM; Phil M. Wagner, "When at Spanish Fort: Amusing Reminiscences of the Time Just Before Evacuation," *National Tribune*, February 24, 1898.

23. Moore to mother, April 10, 1865, Elias Moore Letters, http://www.ohiocivilwar.com/; Scott, comp., *Thirty-Second Iowa*, 336; Tarrant, "Siege and Capture of Fort Blakely," 457–58; clipping from *Evening Gazette* (Burlington, Iowa), November 26, 1909, in A. V. Kendrick Reminiscences, MHM; Sniffen to mother, March [April] 8, 1865, William B. Sniffen Papers, USAMHI; Hart to wife, April 10, 1865, Henry W. Hart Letters, EU; Hills, "Last Battle of the War," 181; Allen, "Some Account and Recollections," 86; Wilson, "The Campaign Ending with the Capture of Mobile," 25; Woods, *Services of the Ninety-Sixth Ohio*, 118; Winschel, ed., *Civil War Diary of a Common Soldier*, 150; Henry Fike to Cimbaline, April 10, 1865, Henry and Lucy Fike Papers, UK; copy of T. J. Stow to brother, April 9, 1865, 37th Illinois Folder, VNMP; Snow to brother, April 14, 1865, David Basset Snow Collection, MU; Taylor Vaughan, "A Johnny's View: Gives a Few Points About Blakeley as He Knew Them," *National Tribune*, February 22, 1900.

24. Marshall, *Eighty-Third Ohio*, 166; Andrews, *History of the Campaign of Mobile*, 202–03; Hart to wife, April 10, 1865, Henry W. Hart Letters, EU; W. R. Eddington, "My Civil War Memoirs and Other Reminiscences," http://macoupinctygenealogy.org/war/edding.html.

25. Snow to brother, April 10, 14, 1865, David Basset Snow Collection, MU; Thomas B. Marshall Diaries, April 9, 1865, MU; Marshall, *History of the Eighty-Third Ohio*, 166; Hatch, ed., *Dearest Susie*, 113.

26. Untitled and incomplete draft of an essay, Lorin Low Dame Correspondence and Journals, MHS; Job H. Yaggy Diary, April 9, 1865, ALPL; Winter, ed., *Captain Joseph Boyce*, 220; Cyrus E. Smith, "Capturing Fort Blakely," *National Tribune*, February 3, 1910.

27. Moore to mother, April 10, 1865, "Elias Moore Letters," http://www.ohiocivilwar.com/; W. R. Eddington, "My Civil War Memoirs and Other Reminiscences," http://macoupinctygenealogy.org/war/edding.html.

28. Tarrant, "Siege and Capture of Fort Blakely," 458; *OR*, vol. 49, pt. 1, 295.

29. O'Brien, *Mobile, 1865*, 202; Wilson, "The Campaign Ending with the Capture of Mobile," 24; Andrew J. Minnick to parents, April 13, 1865, Federal Soldiers Letters, UNC; "Diary of John S. Morgan," 588; Popchock, ed., *Soldier Boy*, 201; Dame to Belle, April 9, 1865, Lorin Low Dame Correspondence and Journals, MHS; Winschel, ed., *Civil War Diary of a Common Soldier*, 150; copy of unidentified diary, April 10, 1865, 130th Illinois Folder, VNMP.

30. Sperry, *History of the 33rd Iowa*, 155; Snow to brother, April 14, 1865, David Basset Snow Collection, MU; Jackson, ed., *"Some of the Boys,"* 246; Moore to mother, April 10, 1865, Elias Moore Letters, http://www.ohiocivilwar.com/; E. A. Stoneburner, "The Siege of Blakely: Comrade Stoneburner Writes of His Experiences With the 114th Ohio," *National Tribune*, November 30, 1899.

31. Moore to mother, April 10, 1865, Elias Moore Letters, http://www.ohiocivilwar.com/.

32. Hart to wife, April 10, 1865, Henry W. Hart Letters, EU; Sniffen to mother, March [April], 8, 1865, William B. Sniffen Papers, USAMHI; diary, April 9, 1865, William J. Gould Papers, LC.

33. W. R. Eddington, "My Civil War Memoirs and Other Reminiscences," http://macoupinctygenealogy.org/war/edding.html.

34. *OR*, vol. 49, pt. 1, 291; Andrews, *History of the Campaign of Mobile*, 201.

35. Hart to wife, April 10, 1865, Henry W. Hart Letters, EU.

36. Woods, *Services of the Ninety-Sixth Ohio*, 122–23.

37. Andrews, *History of the Campaign of Mobile*, 210, 221–22; Jackson, ed., *"Some of the Boys,"* 246; Hart to wife, April 10, 1865, Henry W. Hart Letters, EU.

38. W. R. Eddington, "My Civil War Memoirs and Other Reminiscences," http://macoupinctygenealogy.org/war/edding.html; copy of W. R. Mallory diary, April 9, 1865, 72nd Ohio Folder, VNMP.

39. Jackson, ed., *"Some of the Boys,"* 245–46.

40. Snow to brother, April 14, 1865, David Basset Snow Collection, MU.

41. Moses A. Cleveland Journal, April 9, 1865, MHS.

42. Winter, ed., *Captain Joseph Boyce*, 221.

43. Christensen to Hawkins and Andrews, April 10, 1865, Frederick Steele Papers, SU; Holbrook, *Narrative*, 193; Johnson, *Muskets and Medicine*, 219; Sperry, *History of the 33rd Iowa*, 155; Moses A. Cleveland Journal, April 10, 1865, MHS; Hart to wife, April 10, 1865, Henry W. Hart Letters, EU; Taylor Vaughan, "A Johnny's View: Gives a Few Points about Blakeley as He Knew Them," *National Tribune*, February 22, 1900.

44. Marshall, *History of the Eighty-Third Ohio*, 166; Woods, *Services of the Ninety-Sixth Ohio*, 124.

45. Hiram Scofield Diary, April 10, 1865, NL; copy of unidentified diary, April 11, 1865, 130th Illinois Folder, VNMP; copy of T. J. Stow to brother, April 9, 1865, 37th Illinois Folder, VNMP; Thomas B. Marshall Diaries, April 10, 1865, MU; Jackson, ed., *"Some of the Boys,"* 246; Holbrook, *Narrative*, 193; "Diary of John S. Morgan," 588; Hart to wife, April 10, 1865, Henry W. Hart Letters, EU.

46. Hart to wife, April 10, 1865, Henry W. Hart Letters, EU; Moses S. Cleveland Journal, April 14, 1865, MHS.

47. Allen to Delafield, October 3, 1865, quoted in King, *Torpedoes*, 4; Wilson, "The Campaign Ending with the Capture of Mobile," 17; George W. Read, "Gen. Osterhaus Was Cool," *National Tribune*, September 15, 1904; Sperry, *History of the 33rd Iowa*, 155.

48. Moore to mother, April 10, 1865, Elias Moore Letters, http://www.ohiocivilwar.com/; Sniffen to mother, March [April] 8, 1865 (continued after that date), William B. Sniffen Papers, USAMHI; Dame to Belle, April 9, 1865, Lorin Low Dame Correspondence and Journals, MHS; Woods, *Services of the Ninety-Sixth Ohio*, 118; diary, April 2, 1865, Monroe Joshua Miller Papers, MHM; Jackson, ed., *"Some of the Boys,"* 246; Hart to wife, April 10, 1865, Henry W. Hart Letters, EU.

49. Evans to parents, April 10, 1865, Thomas Evans Letters and Diary, EWU.

50. Moore to mother, April 10, 1865, Elias Moore Letters, http://www.ohiocivilwar.com/; Scott, comp., *Thirty-Second Iowa*, 339.

51. Allen to Delafield, October 3, 1865, quoted in King, *Torpedoes*, 4–5; Jones, *Artillery Fuses*, 127.

52. Allen to Delafield, October 3, 1865, quoted in King, *Torpedoes*, 5; Jones, *Artillery Fuses*, 128.

53. Wilson, "The Campaign Ending with the Capture of Mobile," 18; Delafield, *Report on the Art of War*, 109–110; Scott, comp., *Thirty-Second Iowa*, 339.

54. Diary, April 10, 1865, William J. Gould Papers, LC; Scott, comp., *Thirty-Second Iowa*, 339; Hearn, *Mobile Bay*, 200.

55. Andrews, *History of the Campaign of Mobile*, 160.

56. Diary, April 10, 1865, William J. Gould Papers, LC.

57. *OR*, vol. 49, pt. 2, 197; Elder, ed., *A Damned Iowa Greyhound*, 163–64.

58. Allen to Delafield, October 3, 1865, quoted in King, *Torpedoes*, 4; "Diary of John S. Morgan," 581.

59. Schneck, "Origins of Military Mines: Part I," 53; Hart to wife, April 10, 1865, Henry W. Hart Letters, EU; Snow to brother, April 14, 1865, David Basset Snow Collection, MU.

CHAPTER 10

1. King to Delafield, June 27, 1864, quoted in King, *Torpedoes*, 29–30.

2. Letter to *Newburgh Journal*, February 10, 1864; dispatch, May 27, 1864, *New York Herald*: George H. Hart dispatch, May 26, 1864; C. C. Cusick to editor of *Niagara County Intelligencer*, May 29, 1864. All of these can be found at 132nd Regiment, New York Volunteers Infantry, Civil War Newspaper Clippings, http://dmna.state.ny.us/historic/reghist/civil/infantry/132ndInf/132ndInfCWN.htm; *Chattanooga Daily Rebel*, June 27, 1864.

3. Dispatch, May 27,1864, *New York Herald*; C. C. Cusick to editor of *Niagara County Intelligencer*, May 29, 1864; George H. Hart dispatch, May 26, 1864. All of these can be found at 132nd Regiment, New York Volunteers Infantry, Civil War Newspaper Clippings, http://dmna.state.ny.us/historic/reghist/civil/infantry/132ndInf/132ndInfCWN.htm; Raleigh, North Carolina, *The Confederate*, June 4, 1864.

4. C. C. Cusick to editor of *Niagara County Intelligencer*, May 29, 1864; George H. Hart dispatch, May 26, 1864. All of these can be found at 132nd Regiment, New York Volunteers Infantry, Civil War Newspaper Clippings, http://dmna.state.ny.us/historic/reghist/civil/infantry/132ndInf/132ndInfCWN.htm.

5. *OR*, vol. 36, pt. 3, 245.

6. *OR*, vol. 36, pt. 3, 267.

7. *SOR*, pt. 1, vol. 10, 584.

8. King, *Torpedoes*, 30–32.

9. *OR*, ser. 3, vol. 2, 709.

10. *OR*, ser. 3, vol. 2, 708–09.

11. *OR*, ser. 3, vol. 2, 709–10; Haupt, *Military Bridges*, 124–25; Haupt, *Reminiscences*, 101, 186.

12. Haupt to Julius Moore, November 26, 1862, Herman Haupt Letter Book, Lewis Muhlenberg Haupt Family Papers, LC; Haupt, *Reminiscences*, 186.

13. *OR*, vol. 21, 827.

14. William Fawcett, "Stoneman's Raiders," *National Tribune*, April 4, 1892; *OR*, vol. 25, pt. 1, 1070, 1091.

15. *OR*, vol. 40, pt. 2, 270; *OR*, vol. 40, pt.1, 625, 630. A secondhand reference to torpedoes planted to blow up the railroad bridge at Rappahannock Station appears in the private correspondence of a Sanitary Commission worker. See Samuel W. Richards to Dennett, September 3, 1863, Civil War Letters to William S. Dennett, UVA.

16. Suter to Delafield, October 26, 1864, quoted in King, *Torpedoes*, 32.

17. Suter to Delafield, October 26, 1864, quoted in King, *Torpedoes*, 32–33.

18. Suter to Delafield, October 26, 1864, quoted in King, *Torpedoes*, 32–33.

19. *OR*, vol. 35, pt. 2, 188.

20. *OR*, vol. 35, pt. 1, 410, 419–20; *OR*, vol. 35, pt. 2, 35, 189.

21. Suter to Delafield, October 26, 1864, quoted in King, *Torpedoes*, 32.

22. Doubleday to Foster, July 16, 1864; Foster endorsement, July 23, 1864; and cards. All are found in Charles F. Smith Service Record, Third US Colored Infantry, NARA.

23. *OR*, vol. 42, pt. 1, 743–44.

24. *OR*, vol. 42, pt. 1, 743–44.
25. *OR*, vol. 38, pt. 1, 81; Sherman, *Memoirs*, vol. 2, 105; Harwell and Racine, eds., *Fiery Trail*, 11; diary, August 29, 1864, Abraham J. Seay Collection, UO.
26. *OR*, ser. 2, vol. 8, 49–50.
27. Haupt, *Reminiscences*, 49.
28. Ragan, *Confederate Saboteurs*, 1, 50.
29. Ragan, *Confederate Saboteurs*, 64–65.
30. Wideman, *Sinking of the USS Cairo*, 50–51; *Chattanooga Daily Rebel*, quoted in *Savannah Republican*, July 7, 1863.
31. *OR*, vol. 31, pt. 3, 579; *OR*, vol. 31 pt. 1, 755; Lyon, comp., *Reminiscences*, 131.
32. Fontaine, *My Life*, 212.
33. Fontaine, *My Life*, 212–14; *OR*, vol. 31, pt. 3, 412; Kochan and Wideman, *Civil War Torpedoes*, 311, 658n–59n.
34. *OR*, vol. 29, pt. 2, 652.
35. *OR*, vol. 29, pt. 2, 653.
36. *OR*, vol. 29, pt. 2, 654; Miers, ed., *Rebel War Clerk's Diary*, 269.
37. *OR*, vol. 31, pt. 3, 474; *OR*, vol. 32, pt. 2, 74; Peter J. Osterhaus diary, February 25–26, 1864, Osterhaus Family Papers, MHM.
38. Oladowski to Wright, March 24, 1864, Hypolite Oladowski Service Record, M331, NARA.
39. *Memphis Daily Appeal*, May 3, 1864.
40. Wheeler to Bragg, July, n.d., 1864, Braxton Bragg Papers, DU; *OR*, vol. 38, pt. 2, 865.
41. *OR*, vol. 38 pt. 4, 492, 579.
42. *OR*, vol. 38, pt. 4, 634.
43. Wright, *History of the Eighth Regiment*, 252.
44. Barton, *Autobiography*, 186–87; *OR*, vol. 38, pt. 5, 64.
45. *OR*, vol. 22, pt. 2, 973–74.
46. *OR*, vol. 22, pt. 2, 1072; Hess, *Civil War Logistics*, 226–30.
47. Shannon, "'Infernal Machines' Described," 458.
48. Shannon, "'Infernal Machines' Described," 458.
49. Shannon, "'Infernal Machines' Described," 458.
50. *Macon Daily Telegraph*, March 31, 1864; *Scientific American* 10, issue 17 (April 23, 1864): 272.
51. General Orders No. 194, Headquarters, Mississippi Squadron, May 20, 1864, in Walke, *Naval Scenes*, 453; Porter, "Torpedo Warfare," 225.
52. General Orders No. 194, Headquarters, Mississippi Squadron, May 20, 1864, in Walke, *Naval Scenes*, 453.
53. Tidwell, Hall, and Gaddy, *Come Retribution*, 162; Kochan and Wideman, *Civil War Torpedoes*, 287, 297.
54. Courtenay to Clark, January 19, 1864, quoted in Barnes, *Submarine Warfare*, 76n; Ragan, *Confederate Saboteurs*, 139; *ORN*, vol. 26, 184.
55. *OR*, vol. 20, pt. 1, 758; *OR*, vol. 30, pt. 2, 306; Terry, "An Anecdote," 638.
56. Terry, "An Anecdote," 638.
57. Porter, "Torpedo Warfare," 226; Schiller, ed., *Confederate Torpedoes*, 158.

58. Porter, "Torpedo Warfare," 226; *New York Times*, June 4, 1864; Barnes, *Submarine Warfare*, 75.
59. Barnes, *Submarine Warfare*, 76.
60. *St. Louis Globe Democrat*, May 6, 1888.
61. Salecker, *Disaster on the Mississippi*, 57–59, 65–66, 80–81, 206, 216.
62. *OR*, vol. 46, pt. 3, 1091; Barnes, *Submarine Warfare*, 76. Two operatives disguised as women tried to damage the Springfield Armory in Massachusetts with a coal torpedo. They were discovered by a watchman, who ordered them away, but they left it in the stairwell of the tower, and it was found by another watchman the next morning. Resembling a lump of coal, it contained half a pound of powder inside. See Hunter, "'Patriots,' 'Cowards,' and 'Men Disloyal at Heart,'" 73.
63. Tidwell, Hall, and Gaddy, *Come Retribution*, 162–63; *OR*, vol. 42, pt. 1, 954, 956; King, *Torpedoes*, 7, 16.
64. *OR*, vol. 42, pt. 1, 954–55.
65. *New York Daily Tribune*, August 13, 1864; Schaff, "Explosion at City Point," 481, 485; *OR*, vol. 42, pt. 1, 955.
66. *OR*, vol. 42, pt. 1, 17; Crawford to Samuel P., August 16, 1864, Lewis Crawford Papers, NC; *New York Daily Tribune*, August 13, 1864; Charles Porter, "Explosion at City Point," *National Tribune*, January 7, 1904.
67. *OR*, vol. 42, pt. 1, 17, 955; *New York Daily Tribune*, August 13, 1864; Schaff, "Explosion at City Point," 481–82.
68. Schaff, "Explosion at City Point," 478, 480–81; *OR*, vol. 42, pt. 1, 17, 955.
69. *New York Daily Tribune*, August 13, 1864.
70. *New York Daily Tribune*, August 13, 1864.
71. *New York Daily Tribune*, August 13, 1864; Crawford to Samuel P., August 16, 1864, Lewis Crawford Papers, NC; Schaff, "Explosion at City Point," 483.
72. *OR*, vol. 42, pt. 1, 17, 955; Schaff, "Explosion at City Point," 485; *OR*, vol. 46, pt. 3, 1250.
73. Schaff, "Explosion at City Point," 485.
74. Barnes, *Submarine Warfare*, 75.

CHAPTER 11

1. Schiffer, *Power Struggles*, 130.
2. *SOR*, pt. 3, vol. 2, 388.
3. Rains, "Torpedo Book," in Schiller, ed., *Confederate Torpedoes*, 16.
4. Schiffer, *Power Struggles*, 88–89; *Index of Patents Relating to Electricity*, 6.
5. *Index of Patents Relating to Electricity*, 8; Kochan and Wideman, *Civil War Torpedoes*, 299, 343, 346.
6. Kochan and Wideman, *Civil War Torpedoes*, 299, 346; Schiffer, *Power Struggles*, 167; Plum, *Military Telegraph*, vol. 2, 88–98.
7. Delafield, *Report on the Art of War*, 111–12; King, *Torpedoes*, 53–54, 56–58.
8. King, *Torpedoes*, 56.

9. Schiffer, *Power Struggles*, 232; Simon, ed., *Papers of Ulysses S. Grant*, vol. 14, 72n; "Taliferro P. Shaffner," https://en.wikipedia.org/wiki/Taliaferro_Preston_Shaffner.

10. Simon, ed., *Papers of Ulysses S. Grant*, vol. 14, 70–71, 71n–72n.

11. Simon, ed., *Papers of Ulysses S. Grant*, vol. 14, 72n–73n.

12. *OR*, vol. 46, pt. 1, 172; Simon, ed., *Papers of Ulysses S. Grant*, vol. 14, 72n.

13. Simon, ed., *Papers of Ulysses S. Grant*, vol. 14, 72n.

14. Simon, ed., *Papers of Ulysses S. Grant*, vol. 14, 73n.

15. "Experiments with the American Torpedo-Shells," 357.

16. "Experiments with the American Torpedo-Shells," 357.

17. "Experiments with the American Torpedo-Shells," 357–58.

18. Allen Mesch, "Colonel Edward Roberts," http://salientpoints.blogspot.com/2014/04/colonel.edward.roberts.html; Whiteshot, *Oil-Well Driller*, 754–755.

19. Whiteshot, *Oil-Well Driller*, 755–56; Allen Mesch, "Colonel Edward Roberts," http://salientpoints.blogspot.com/2014/04/colonel.edward.roberts.html.

20. Whiteshot, *Oil-Well Driller*, 756; *Titusville Morning Herald*, June 17, 1866, October 28, 1868; "Torpedo (petroleum)," https://en.wikipedia.org/wiki/Torpedo_(petroleum).

21. "Torpedo (petroleum)," https://en.wikipedia.org/wiki/Torpedo_(petroleum); "Hydraulic fracturing," https://en.wikipedia.org/wiki/Hydraulic_fracturing.

22. Undated, unidentified newspaper clipping by P. Pitkin, Francis W. Tupper Papers, ALPL.

23. Beauregard to Dear Sir, February 15, 1876, *Journal* (Kingston, New York), March 20, 1878.

24. Hays, comp., *Under the Red Patch*, 84.

25. Waters, "'Deception Is the Art of War,'" 57; Schiller, ed., *Confederate Torpedoes*, 7.

26. Rains, "Torpedo Book," in Schiller, ed., *Confederate Torpedoes*, 55, 59.

27. Rains, "Torpedo Book," in Schiller, ed., *Confederate Torpedoes*, 56.

28. Rains, "Torpedo Book," in Schiller, ed., *Confederate Torpedoes*, 56, 63.

29. Rains, "Torpedo Book," in Schiller, ed., *Confederate Torpedoes*, 63.

30. Rains, "Torpedo Book," in Schiller, ed., *Confederate Torpedoes*, 54.

31. Rains, "Torpedo Book," in Schiller, ed., *Confederate Torpedoes*, 58–59, 60, 63.

32. Rains, "Torpedo Book," in Schiller, ed., *Confederate Torpedoes*, 58, 69–73.

33. Simon, ed., *Papers of Ulysses S. Grant*, vol. 25, 422–23.

34. Simon, ed., *Papers of Ulysses S. Grant*, vol. 25, 423.

35. Ernst, *Manual for Practical Military Engineering*, 224.

36. Rains, "Torpedoes," 255–56, 260.

37. Rains, "Torpedoes," 257–60.

38. Rains to Walthall, June 21, 1879, in Waters and Brown, *Gabriel Rains*, 128, 130, 132.

39. Davis, *Rise and Fall*, vol. 2, 97–98, 207–08.

40. "Invention of Torpedoes," 734; Waters, "'Deception Is the Art of War,'" 57, 59; *National Cyclopaedia*, vol. 4, 336; Schiller, ed., *Confederate Torpedoes*, 7–8.

41. Davidson, "Electrical Submarine Mine," 459.

42. Davidson, "Electrical Submarine Mine," 456; Maury, "Defence of Spanish Fort," 134; Rains, "Torpedoes," 256; Porter, "Torpedo Warfare," 228; Schiffer, *Power Struggles*, 127, 129; Perry, *Infernal Machines*, 199–201.

43. *The Farmer and Mechanic* (Raleigh, North Carolina), April 25, 1883; Alexander, "Sketch of Longstreet's Division," 39n.

44. Kochan and Wideman, *Civil War Torpedoes*, 231.

45. Johnston, "Manassas to Seven Pines," 205, 212n; "Confederate Use of Subterranean Shells," 201.

46. Perry, *Infernal Machines*, 194; Youngblood, *Development of Mine Warfare*, 65; Campbell, ed. *Engineer in Gray*, 133–58.

47. John W. Burke Memoir, WLC-UM; Adams, *Trooper's Adventures*, 382; Marks, *Peninsular Campaign*, 151; Banes, *History of the Philadelphia Brigade*, 56; Floyd, *History of the Fortieth (Mozart) Regiment*, 142; Hyde, *Following the Greek Cross*, 48; William Simmons, "Confederate Torpedoes," *National Tribune*, April 28, 1892.

48. Michie, "Notes Explaining Rebel Torpedoes and Ordnance," in Schiller, ed., *Confederate Torpedoes*, 95, 114–15, 117–18, 129–30.

49. King, *Torpedoes*, frontispiece; Wolters, "Electric Torpedoes," 758.

50. King, *Torpedoes*, 31, 86–88.

51. King, *Torpedoes*, 72–73, 87–88.

52. King, *Torpedoes*, 86; "Guncotton," https://en.wikipedia.org/wiki/Nitrocellulose.

53. Youngblood, *Development of Mine Warfare*, 72.

54. Barnes, *Submarine Warfare*, 16, 63.

55. Barnes, *Submarine Warfare*, 62–63.

56. Wolters, "Electric Torpedoes," 760; Schiffer, *Power Struggles*, 130; Porter, "Torpedo Warfare," 233.

57. Porter, "Torpedo Warfare," 213–14, 230.

58. Trowbridge, *Desolate South*, 108.

59. Sperry, *History of the 33rd Iowa*, 155.

60. Allen to Delafield, October 3, 1865, quoted in King, *Torpedoes*, 5.

61. Schneck, "Origins of Military Mines: Part I," 52; Lord, *Civil War Collector's Encyclopedia*, 173; Croll, *History of Landmines*, 20; Schiller, ed., *Confederate Torpedoes*, 169–70. Historian Chester G. Hearn has stated that civilians were killed and injured by landmines at Mobile for decades after the war but offers no proof of any kind to support the assertion. See Hearn, *Mobile Bay*, 200.

62. Allen to Delafield, October 3, 1865, quoted in King, *Torpedoes*, 4.

CHAPTER 12

1. Gray, *Nineteenth-Century Torpedoes*, 5, 8; Youngblood, *Development of Mine Warfare*, 73; "Torpedo," https://en.wikipedia.org/wiki/torpedo.

2. Sater, *Andean Tragedy*, 18–21, 250, 252.

3. Sater, *Andean Tragedy*, 288.

4. Croll, *History of Landmines*, 13, 20–21; Youngblood, *Development of Mine Warfare*, 73.

5. Toepfer, "Technics in the Russo-Japanese War," 197; "German Regulations," 367, 369; James, *Siege of Port Arthur*, 84; Youngblood, *Development of Mine Warfare*, 74–78; Abbott and Cassedy, "Land Mines," 367; Schneck, "Origins of Military Mines: Part II," 52.

6. Youngblood, *Development of Mine Warfare*, 60, 82.

7. Youngblood, *Development of Mine Warfare*, 80–82.

8. East, ed., *Gallipoli Diary*, 125.

9. Abbott and Cassedy, "Land Mines," 367; Schneck, "Origins of Military Mines: Part II," 44; Youngblood, *Development of Mine Warfare*, 103, 105; Jones, *Malice Aforethought*, 16; Trounce, *Fighting the Boche Underground*, 154–55; Ball, "Work of the Miner," 238–39; Grieve and Newman, *Tunnellers*, 145, 148.

10. Graham, *Life of a Tunnelling Company*, 162; Ball, "Work of the Miner," 232, 235–36.

11. Croll, *History of Landmines*, 25, 27, 29, 31.

12. Youngblood, *Development of Mine Warfare*, 106–07.

13. *Manual of Field Works*, 196–98; Youngblood, *Development of Mine Warfare*, 107.

14. Abbott and Cassedy, "Land Mines," 367–68; Croll, *History of Landmines*, 35–36.

15. Youngblood, *Development of Mine Warfare*, 107; Schneck, "Origins of Military Mines: Part I," 52; Abbott and Cassedy, "Land Mines," 368; Croll, *History of Landmines*, 37.

16. Abbott and Cassedy, "Land Mines," 368.

17. Croll, *History of Landmines*, 38.

18. Croll, *History of Landmines*, 42–43, 73; Abbott and Cassedy, "Land Mines," 367–68; Schneck, "Origins of Military Mines: Part II," 45; Kochan and Wideman, *Civil War Torpedoes*, 298; Bull and Rottman, *Infantry Tactics*, 117; Rottman, *World War II Axis Booby Traps*, 9.

19. Croll, *History of Landmines*, 39; Jones, *Malice Aforethought*, 14; Bull and Rottman, *Infantry Tactics*, 117, 119–21.

20. Croll, *History of Landmines*, 41–42, 71–73; Rottman, *World War II Axis Booby Traps*, 62–63; Doubler, *Closing with the Enemy*, 26, 112; Jones, *Malice Aforethought*, 12.

21. Croll, *History of Landmines*, 67–68, 71; Doubler, *Closing with the Enemy*, 138.

22. Croll, *History of Landmines*, 53–54, 56, 65–66; Youngblood, *Development of Mine Warfare*, 139.

23. Croll, *History of Landmines*, 88–89.

24. Croll, *History of Landmines*, 91–93; Youngblood, *Development of Mine Warfare*, 139–40.

25. Schneck, "Origins of Military Mines: Part I," 54; Abbott and Cassedy, "Land Mines," 368; Croll, *History of Landmines*, 14.

26. Croll, *History of Landmines*, 110–12.

27. Croll, *History of Landmines*, 96–97, 104, 106–07.

28. Croll, *History of Landmines*, 102, 118–20, 122.

29. Croll, *History of Landmines*, 128–33.

30. Croll, *History of Landmines*, 133–34.

31. Jones, *Malice Aforethought*, 13; "Ottawa Treaty," https://en.wikipedia.org/wiki/Ottawa_Treaty; Efaw, "United States Refusal," 131–32, 140; Croll, *History of Landmines*, 135; Bower, "Norms Without the Great Powers," 355; Sigal, *Negotiating Minefields*, 3.

32. Efaw, "United States Refusal," 143–44, 149–50; Bower, "Norms Without the Great Powers," 356.

33. Efaw, "United States Refusal," 93–94, 95n, 106.

34. Bower, "Norms Without the Great Powers," 356.

35. Bower, "Norms Without the Great Powers," 357. For a good overview of the crusade to ban landmines, see Williams, Goose, and Wareham, eds., *Banning Landmines*.

36. *The Journal of ERW and Mine Action*, http://www.jmu.edu/(isr/journal/18.3/index.shtml.

CONCLUSION

1. Abbott and Cassedy, "Land Mines—Past and Present," 367; Perry, *Infernal Machines*, 195, 197; Robson, "Star-Spangled Land Mines," 354; Waters, "'Deception Is the Art of War,'" 60; Croll, *History of Landmines*, 10; Youngblood, *Development of Mine Warfare*, 35, 60.

2. Croll, *History of Landmines*, 18.

3. Croll, *History of Landmines*, 16.

4. Richter, *Chemical Soldiers: British Gas Warfare*, 1, 8, 10, 17.

5. Friedel, *Francis Lieber: Nineteenth-Century Liberal*, 2, 15–16, 28, 52, 82, 122, 293, 306, 324–26, 328, 331–35.

6. *OR*, ser. 3, vol. 3, 148–64; Friedel, *Francis Lieber*, 325. Lieber did not mention torpedoes in any other writings, including *Contributions to Political Science* and *Reminiscences, Addresses, and Essays*.

7. Friedel, *Francis Lieber*, 337, 339–40; see also Witt, *Lincoln's Code: The Laws of War*, passim.

8. *OR*, vol. 11, pt. 1, 349–50.

9. *OR*, vol. 11, pt. 1, 350.

10. *OR*, vol. 11, pt. 1, 349.

11. *OR*, vol. 11, pt. 1, 349–50.

APPENDIX

1. Kochan and Wideman, *Civil War Torpedoes*, 260, 262.

2. Read, "Gen. Osterhaus Was Cool," *National Tribune*, September 15, 1904.

3. "List of Severe wounds in 2d Div. 15th A.C. rec'd at Statesboro Ga. Dec. 4th & in assault on Fort McAllister, Dec. 13th 1864, 'Sherman's March to the Sea,'" (folder 457, series 1, roll 3), United States Sanitary Commission Records, NYPL.

4. "Samuel Weissell Gross," https://snaccooperative.org/ark:/99166/w6g7555s.

5. Gross, "On Torpedo Wounds," 370–71.

6. Gross, "On Torpedo Wounds," 371.

7. Gross, "On Torpedo Wounds," 371.

8. Gross, "On Torpedo Wounds," 371–72.

9. Gross, "On Torpedo Wounds," 372.

10. Gross, "On Torpedo Wounds," 372.

11. Jack Bryant Stinson service record, Forty-first Alabama, M311, NARA; Jerry Burkhalter service record, Forty-first Alabama, M311, NARA; Stimson, "Three Unusual Gun-Shot Wounds," 364.

12. Stimson, "Three Unusual Gun-Shot Wounds," 364; Kochan and Wideman, *Civil War Torpedoes*, 267.

Bibliography

ARCHIVES

Abraham Lincoln Presidential Library, Springfield, Illinois
   James F. Drish Papers
   Francis W. Tupper Papers
   Job H. Yaggy Diary
Bowdoin College, Special Collections and Archives, Brunswick, Maine
   O. O. Howard Papers
Cheshire Archives and Local Studies, Chester, England
   Letters and Papers of Colonel Hugh Robert Hibbert
Chicago History Museum, Chicago, Illinois
   George Carrington Diary
   Spencer S. Kimbell Papers
   John Merrilles Diary
Cincinnati History Museum, Cincinnati, Ohio
   Fox Collection
College of William and Mary, Special Collections, Williamsburg, Virginia
   Bowles Family Papers
Duke University, Rubenstein Rare Book and Manuscript Library, Durham, North Carolina
   Braxton Bragg Papers
   Bell Halsey Miller Papers
   Thomas Sewell Papers
   John C. Van Duzer Diary
Eastern Washington University, Archives and Special Collections, Cheney
   Thomas Evans Letters and Diary
Emory University, Manuscript, Archives, and Rare Book Library, Atlanta, Georgia
   Henry W. Hart Letters
   Alfred A. Rigby Diary
Filson Historical Society, Louisville, Kentucky
   Schmidt Family Papers
Harvard University, Houghton Library, Cambridge, Massachusetts

Frederick M. Dearborn Collection
Huntington Library, San Marino, California
  Theophilus M. Magaw Papers
  Edward E. Schweitzer Papers
Library of Congress, Manuscript Division, Washington, DC
  G. T. Beauregard Papers
  C. B. Comstock Papers
  William J. Gould Papers
  Lewis Muhlenburg Haupt Family Papers
  William T. Sherman Papers
Lincoln Memorial University, Abraham Lincoln Library and Museum, Harrogate, Tennessee
  Oliver O. Howard Papers
Louisiana State University, Louisiana and Lower Mississippi Valley Collections, Special Collections, Baton Rouge
  Wayne Johnson Jacobs Diaries and Lists
Mariners' Museum Library, Newport News, Virginia
  William Byrnes Letter
Maryland Historical Society, Baltimore
  Charles E. Phelps "Personal Recollections of the Wilderness Campaign"
Massachusetts Historical Society, Boston
  Moses A. Cleveland Journal
  Lorin Low Dame Correspondence and Journals
Miami University, Walter Havighurst Special Collections, Oxford, Ohio
  Thomas B. Marshall Diaries
  David Basset Snow Collection
Minnesota Historical Society, St. Paul
  George R. Buckman Civil War Papers
Mississippi Department of Archives and History, Jackson
  William D. Elder Papers
Missouri History Museum, St. Louis
  A. V. Kendrick Reminiscences
  Monroe Joshua Miller Papers
  Osterhaus Family Papers
National Archives and Records Administration, Washington, DC
  Jerry Burkhalter Service Record, Forty-first Alabama, M311, Compiled Service Records of Confederate Soldiers Who Served in Organizations from the State of Alabama, RG109
  Marshall McDonald Service Record, M331, Compiled Service Records of Confederate General and Staff Officers and Non-Regimental Enlisted Men, RG109
  Hypolite Oladowski Service Record, M331, Compiled Service Records of Confederate General and Staff Officers and Non-Regimental Enlisted Men, RG109
  Gabriel J. Rains Service Record, Seventh US Infantry, A 130, Adjutant General, Letters Received, RG 94

Gabriel J. Rains Service Record, M331, Compiled Service Records of Confederate General and Staff Officers and Non-Regimental Enlisted Men, RG109

Charles F. Smith Service Record, Third US Colored Infantry, Compiled Service Records of Volunteer Union Soldiers Who Served with United States Colored Troops, RG94

Jack Bryant Stinson Service Record, Forty-first Alabama, M311, Compiled Service Records of Confederate Soldiers Who Served in Organizations from the State of Alabama, RG109

Francis W. Tupper Pension Record, First Alabama Cavalry (US)

Francis W. Tupper Service Record, First Alabama Cavalry (US), M276, Compiled Service Records of Volunteer Union Soldiers Who Served in Organizations from the State of Alabama

Letters and Telegrams Sent by the Engineer Bureau of the Confederate War Department, M628, RG 109

Navarro College, Pearce Museum, Corsicana, Texas
- Lewis Crawford Papers
- George Wheeler Papers

Newberry Library, Chicago, Illinois
- Curtis P. Lacey Diary
- Hiram Scofield Diary

New York Public Library, Rare Books and Manuscripts, New York
- United States Sanitary Commission Records

Ohio Historical Society, Columbus
- Hugh Boyd Ewing Papers

Richmond National Battlefield, Richmond, Virginia
- Clifford Dickinson "Union and Confederate Engineering Operations at Chaffin's Bluff/Chaffin's Farm, June 1862–April 3, 1865"

South Carolina Historical Society, Charleston
- Gabriel J. Rains Papers

Southern Illinois University, Special Collections Research Center, Carbondale
- Gabriel J. Rains Letter, Civil War Collection

Stanford University, Special Collections, Stanford, California
- Frederick Steele Papers

Tennessee State Library and Archives, Nashville
- Thomas B. Beggs Diary, Civil War Collection

Tulane University, Special Collections, New Orleans, Louisiana
- Battalion Washington Artillery Collection, Civil War Papers, Louisiana Historical Association Collection
- George W. Brent Papers

US Army Military History Institute, Carlisle, Pennsylvania
- Carl W. Bernhardt Diary, Bernhardt-Campbell Family Papers
- Gordon Hickenlooper, ed. "The Reminiscences of General Andrew Hickenlooper, 1861–1865," *Civil War Times* Collection, Series 2
- William B. Sniffen Papers
- James M. Uhler Papers

University of Georgia, Hargrett Rare Book and Manuscript Library, Athens
  William Starr Basinger Reminiscences
University of Kansas, Spencer Research Library, Lawrence
  Henry and Lucy Fike Papers
University of Michigan, Bentley Historical Library, Ann Arbor
  William E. Sleight Reminiscences
University of Michigan, William L. Clements Library, Ann Arbor
  John W. Burke Memoir, James M. Schoff Civil War Collections
  Calvin D. Mehaffey Papers, James M. Schoff Civil War Collections
University of North Carolina, Southern Historical Collection, Chapel Hill
  Federal Soldiers' Letters
  Samuel H. Lockett Papers
University of Notre Dame, Rare Books and Special Collections, South Bend, Indiana
  Caley Family Correspondence
University of Oklahoma, Western History Collection, Norman
  Abraham J. Seay Collection
University of the South, Archives and Special Collections, Sewanee, Tennessee
  Leonidas Polk Papers
University of South Carolina, South Caroliniana Library, Columbia
  G. T. Beauregard Papers
University of Virginia, Special Collections, Charlottesville
  Civil War Letters to William S. Dennett
  Minor Family Papers
Vicksburg National Military Park, Vicksburg, Mississippi
  37th Illinois Folder
  130th Illinois Folder
  72nd Ohio Folder
Western Reserve Historical Society, Cleveland, Ohio
  Wells A. Bushnell Memoirs
Wisconsin Historical Society, Madison
  John G. Jones Civil War Letters

NEWSPAPERS

The Alleghanian, Ebensburg, Pennsylvania
Daily Comet, Baton Rouge, Louisiana
Belmont Chronicle, St. Clairsville, Ohio
Chattanooga Daily Rebel
Chicago Times
Cincinnati Commercial
Cincinnati Gazette
The Confederate, Raleigh, North Carolina
Daily Argus, Rock Island, Illinois
The Farmer and Mechanic, Raleigh, North Carolina

Frank Leslie's Illustrated Newspaper
Gazette of the United States, Washington, DC
Harper's Weekly
Journal, Kingston, New York
Macon Daily Telegraph, Macon, Georgia
Memphis Daily Appeal
Nashville Union
National Republican, Washington, DC
Newburgh Journal
New York Daily Tribune
New York Herald
New York Times
Niagara County Intelligencer
Rutland Weekly Herald, Rutland, Vermont
St. Louis Daily Missouri Democrat
St. Louis Globe Democrat
Savannah Republican
Scientific American
Titusville Morning Herald
Wisconsin Daily Patriot, Madison, Wisconsin

WEBSITES

"132nd Infantry Regiment." NY Volunteer Infantry Civil War Newspaper Clippings. http://dmna.state.ny.us/historic/reghist/civil/infantry/132ndInf/132ndInfCWN.htm. Accessed July 17, 2022.

"Charles Wheatstone." Wikipedia entry. https://en.wikipedia.org/wiki/Charles_Wheatstone. Accessed July 17, 2022.

"Elias Moore Letters." Ohio in the Civil War. www.ohiocivilwar.com. Accessed July 17, 2022.

"Frederick Augustus Abel." Wikipedia entry. https://en.wikipedia.org/wiki/Frederick_Abel. Accessed July 17, 2022.

"Guncotton." Wikipedia entry. https://en.wikipedia.org/wiki/Nitrocellulose. Accessed July 17, 2022.

"Hydraulic Fracturing." Wikipedia entry. https://en.wikipedia.org/wiki/Hydraulic_fracturing. Accessed July 17, 2022.

"Immanuel Nobel the Younger." Wikipedia entry. https://en.wikipedia.org/wiki/Immanuel_Nobel. Accessed July 17, 2022.

Mesch, Allen. *Colonel Edward Roberts* (blog). *Blogger.* http://salientpoints.blogspot.com/2014/04/colonel-edward.roberts.html. Accessed July 17, 2022.

"Moritz Hermann von Jacobi." Wikipedia entry. https://en.wikipedia.org/wiki/Moritz_von_Jacobi. Accessed July 17, 2022.

"Ottawa Treaty." Wikipedia entry. https://en.wikipedia.org/wiki/Ottawa_Treaty. Accessed July 17, 2022.

"Samuel Weissell Gross." Social Networks and Archival Context (SNAC). https://snaccooperative.org/ark:/99166/w6g7555s. Accessed July 17, 2022.

"Tales of Destruction . . . The Roberts Torpedo." *Titusville Morning Herald*, http://www.logwell.com/tales/roberts_torpedo.html. Accessed July 17, 2022.

"Taliferro P. Shaffner." Wikipedia entry. https://en.wikipedia.org/wiki/Taliaferro_Preston_Shaffner. Accessed July 17, 2022.

"The Journal of ERW and Mine Action." http://www.jmu.edu/(isr/journal/18.3/index.shtml. Accessed July 17, 2022.

"Torpedo." Wikipedia entry. https://en.wikipedia.org/wiki/torpedo. Accessed July 17, 2022.

"Torpedo (Petroleum)." Wikipedia entry. https://en.wikipedia.org/wiki/Torpedo_(petroleum). Accessed July 17, 2022.

"Voltaic Pile." Wikipedia entry. https://en.wikipedia.org/wiki/Voltaic_pile. Accessed July 17, 2022.

"William Robert Grove." Wikipedia entry. https://en.wikipedia.org/wiki/William_Robert_Grove. Accessed July 17, 2022.

W. R. Eddington, "My Civil War Memoirs and Other Reminiscences," http://macoupinctygenealogy.org/war/edding.html. Accessed July 17, 2022.

ARTICLES, BOOKS

Abbot, Henry L. *Siege Artillery in the Campaigns Against Richmond*. New York: D. Van Nostrand, 1868.

Abbott, Jackson M., and Logen Cassedy. "Land Mines—Past and Present." *Military Engineer* 54 (1962): 367–68.

Adams, Francis Colburn. *A Trooper's Adventures in the War for the Union*. New York: Dick and Fitzgerald, 1865.

Aldrich, Thomas M. *The History of Battery A, First Regiment Rhode Island Light Artillery in the War to Preserve the Union, 1861–1865*. Providence, Rhode Island: Snow and Farnham, 1904.

Alexander, E. P. "Sketch of Longstreet's Division—Yorktown and Williamsburg." *Southern Historical Society Papers* 10 (1882): 33–45.

Allen, Charles J. "Some Account and Recollections of the Operations against the City of Mobile and Its Defences, 1864 and 1865." In *Glimpses of the Nation's Struggle: A Series of Papers Read Before the Minnesota Commandery of the Military Order of the Loyal Legion of the United States*. First Series. St. Paul: St. Paul Book, 1887: 54–88.

Ammen, Daniel. "DuPont and the Port Royal Expedition." In *Battles and Leaders of the Civil War*. Vol. 1, edited by Robert Underwood Johnson and Clarence Clough Buel. New York: Thomas Yoseloff, 1956: 671–91.

Andrews, C. C. *History of the Campaign of Mobile*. New York: D. Van Nostrand, 1889.

Ball, H. Standish. "The Work of the Miner on the Western Front." *Transactions of the Institution of Mining and Metallurgy* 28 (1918–1919): 189–85.
Banes, Charles H. *History of the Philadelphia Brigade*. Philadelphia: J. B. Lippincott, 1876.
Barnes, J. S. *Submarine Warfare, Offensive and Defensive*. New York, New York: D. Van Nostrand, 1869.
Barton, T. H. *Autobiography of Dr. Thomas H. Barton, the Self-Made Physician of Syracuse, Ohio*. Charleston: West Virginia Printing, 1890.
Baumgart, Winfried. *The Crimean War, 1853–1856*. 2nd ed. London: Bloomsbury Academic, 2019.
Bell, Jack. *Civil War Heavy Explosive Ordnance: A Guide to Large Artillery Projectiles, Torpedoes, and Mines*. Denton: University of North Texas Press, 2003.
Bentley, Nicolas, ed. *Russell's Dispatches from the Crimea, 1854–1856*. London: Andre Deutsch, 1966.
Bergeron, Arthur W., Jr. *Confederate Mobile*. Jackson: University Press of Mississippi, 1991.
Bombaugh, Charles C. "Extracts from a Journal Kept During the Earlier Campaigns of the Army of the Potomac." *Maryland Historical Magazine* 5, no. 4 (December 1910): 301–26.
Bower, Adam. "Norms Without the Great Powers: International Law, Nested Social Structures, and the Ban on Antipersonnel Mines." *International Studies Review* 17 (2015): 347–73.
Bryan, Charles F., Jr., and Nelson D. Lankford, eds. *Eye of the Storm: A Civil War Odyssey*. New York: Free Press, 2000.
Bull, Stephen, and Gordon L. Rottman. *Infantry Tactics of the Second World War*. Midland House, West Way, Botley, Oxford, United Kingdom: Osprey, 2008.
Campbell, R. Thomas, ed. *Engineer in Gray: Memoirs of Chief Engineer James H. Tomb, CSN*. Jefferson, North Carolina: McFarland, 2005.
Castel, Albert. *Tom Taylor's Civil War*. Lawrence: University Press of Kansas, 2000.
"A Chat With Sheridan: The Story of His Raid Around Richmond, as Told by Himself." *National Tribune*, May 8, 1884.
Cheney, William H. *The Fourteenth Regiment Rhode Island Heavy Artillery (Colored) in the War to Preserve the Union, 1861–1865*. Providence: Snow and Farnham, 1898.
"The Civil War Diary of William M. Macy." *Indiana Magazine of History* 30, no. 2 (June 1934): 181–97.
Clark, George P., ed. "'Reminiscence of My Army Life.'" *Indiana Magazine of History* 101, no. 1 (March 2005): 15–57.
Clark, Olynthus B., ed. *Downing's Civil War Diary, by Sergeant Alexander G. Downing, Company E, Eleventh Iowa Infantry, Third Brigade, 'Crocker's Brigade,' Sixth Division of the Seventeenth Corps, Army of the Tennessee, August 15, 1861–July 31, 1865*. Des Moines, Iowa: Homestead Printing, 1916.
"Confederate Use of Subterranean Shells on the Peninsula." Vol. 2 of *Battles and Leaders of the Civil War*, edited by Robert Underwood Johnson and Clarence Clough Buel. New York: Thomas Yoseloff, 1956: 201.

Connelly, Thomas W. *History of the Seventieth Ohio Regiment*. Cincinnati, Ohio: Peak Brothers, 1902.

Copp, Elbridge J. *Reminiscences of the War of the Rebellion, 1861–1865*. Nashua, New Hampshire: Telegraph Publishing, 1911.

Crist, Lynda Lasswell, ed. *The Papers of Jefferson Davis*. 17 vols. Baton Rouge: Louisiana State University Press, 1971–2008.

Croll, Mike. *The History of Landmines*. Barnsley, United Kingdom: Leo Cooper, 1998.

Crooke, George. *The Twenty-First Regiment of Iowa Volunteer Infantry*. Milwaukee, Wisconsin: King, Fowle, 1891.

Crowell, William C. "With Mulligan at Lexington, Mo." *National Tribune Scrap Book*, no. 2 (n.d.): 44–47.

Crowninshield, Benjamin W. *A History of the First Regiment of Massachusetts Cavalry Volunteers*. Boston: Houghton Mifflin, 1891.

Cutrer, Thomas W. *Theater of a Separate War: The Civil War West of the Mississippi River, 1861–1865*. Chapel Hill: University of North Carolina Press, 2017.

Davidson, Hunter. "The Electrical Submarine Mine—1861–1865." *Confederate Veteran* 16 (1908): 456–59.

Davis, Jefferson. *The Rise and Fall of the Confederate Government*. 2 vols. New York: D. Appleton, 1881.

Davis, Nicholas A. *The Campaign from Texas to Maryland with the Battle of Fredericksburg*. Austin, Texas: Steck, 1961.

Davis, W. W. H. *History of the 104th Pennsylvania Regiment from August 22nd, 1862, to September 30th, 1864*. Philadelphia: James B. Rodgers, 1866.

Delafield, Richard. *Report on the Art of War in Europe in 1854, 1855, and 1856*. Washington, DC: George W. Bowman, 1860.

Denison, Frederic. *Shot and Shell: The Third Rhode Island Heavy Artillery Regiment in the Rebellion, 1861–1865*. Providence: J. A. and R. A. Reid, 1879.

"Diary of John S. Morgan, Company G, Thirty-Third Iowa Infantry." *Annals of Iowa*, Third Series 13, no. 8 (April 1923): 570–610.

Dickey, Luther S. *History of the Eighty-fifth Regiment Pennsylvania Volunteer Infantry, 1861–1865*. New York: J. C. and W. E. Powers, 1915.

Dictionary of American Biography. 10 vols. New York: Charles Scribner's Sons, 1958–1995.

Donald, David Herbert, ed. *Gone for a Soldier: The Civil War Memoirs of Private Alfred Bellard*. Boston: Little, Brown and Company, 1975.

Doubler, Michael D. *Closing with the Enemy: How GIs Fought the War in Europe, 1944–1945*. Lawrence: University Press of Kansas, 1994.

Dudley, Henry Walbridge. *Autobiography*. Menasha, Wisconsin: George Banta, 1914.

East, Ronald, ed. *The Gallipoli Diary of Sergeant Lawrence, of the Australian Engineers—1st A.I.F. 1915*. Carlton, Victoria, Australia: Melbourne University Press, 1981.

Efaw, Andrew C. S. "The United States Refusal to Ban Landmines: The Intersection between Tactics, Strategy, Policy, and International Law." *Military Law Review* 159 (March 1999): 87–151.

Eisenschiml, Otto, ed. *Vermont General: The Unusual War Experiences of Edward Hastings Ripley, 1862–1865.* New York: Devin-Adair, 1960.

Elder, Donald C. III, ed. *A Damned Iowa Greyhound: The Civil War Letters of William Henry Harrison Clayton.* Iowa City: University of Iowa Press, 1998.

Ellison, Janet Correll, ed. *On to Atlanta: The Civil War Diaries of John Hill Ferguson, Illinois Tenth Regiment of Volunteers.* Lincoln: University of Nebraska Press, 2001.

Ernst, O. H. *A Manual for Practical Military Engineering, Prepared for the Use of the Cadets of the US Military Academy, and for Engineer Troops.* New York: D. Van Nostrand, 1873.

"Experiments with the American Torpedo-Shells at Chatham." *Illustrated London News* 47, no. 1338 (October 14, 1865): 357–58.

Fawcett, William. "Stoneman's Raiders." *National Tribune,* April 4, 1892.

Fletcher, Ian, and Natalia Ishchenko. *The Crimean War: A Clash of Empires.* Staplehurst, Kent, United Kingdom: Spellmount, 2004.

Floyd, Fred C. *History of the Fortieth (Mozart) Regiment New York Volunteers.* Boston: F. H. Gilson, 1909.

Fontaine, Lamar. *My Life and My Lectures.* New York: Neale, 1908.

Fonvielle, Chris E., Jr. *The Wilmington Campaign: Last Rays of Departing Hope.* Mechanicsburg, Pennsylvania: Stackpole Books, 1997.

"Fortification and Siege of Port Hudson." *Southern Historical Society Papers* 14 (1886): 305–48.

Foster, Alonzo. *Reminiscences and Record of the 6th New York V. V. Cavalry.* [Brooklyn: n.p.], 1892.

Friedel, Frank. *Francis Lieber: Nineteenth-Century Liberal.* Baton Rouge: Louisiana State University Press, 1948.

"From the Diary of a Private." *New York Times,* June 11, 1893.

Furney, L. A., ed. *Reminiscences of the War of the Rebellion, 1861–1865.* Flushing, New York: Estate of Jacob Roemer, 1897.

Gallagher, Gary W., ed., *Fighting for the Confederacy: The Personal Recollections of General Edward Porter Alexander.* Chapel Hill: University of North Carolina Press, 1989.

"German Regulations for Field Fortifications and Conclusions Reached in Russia from the Battles in Defensive Positions in Manchuria." *Professional Memoirs, Corps of Engineers, United States Army, and Engineer Department at Large* 2, no. 7 (July–September 1910): 347–69.

Gillmore, Quincy A. "The Army before Charleston in 1863." In *Battles and Leaders of the Civil War.* Vol. 4, edited by Robert Underwood Johnson and Clarence Clough Buel. New York: Thomas Yoseloff, 1956: 52–71.

Gordon, George H. *A War Diary of Events in the War of the Great Rebellion.* Boston: James R. Osgood, 1882.

Gould, David, and James B. Kennedy, eds. *Memoirs of a Dutch Mudsill: The 'War Memories' of John Henry Otto, Captain, Company D, 21st Regiment Wisconsin Volunteer Infantry.* Kent, Ohio: Kent State University Press, 2004.

Gracey, S. L. *Annals of the Sixth Pennsylvania Cavalry.* N.p.: E. H. Butler, 1868.

Graham, H. W. *The Life of a Tunnelling Company, Being an Intimate Story of the Life of the 185th Tunnelling Company, Royal Engineers, in France, during the Great War, 1914–1918*. Hexham, England: J. Catherall, 1927.

Gray, Edwyn. *Nineteenth-Century Torpedoes and Their Inventors*. Annapolis, Maryland: Naval Institute Press, 2004.

Grieve, W. Grant, and Bernard Newman. *Tunnellers: The Story of the Tunnelling Companies, Royal Engineers, during the World War*. London: Herbert Jenkins, 1936.

Griffin, Richard N., ed. *Three Years a Soldier: The Diary and Newspaper Correspondence of Private George Perkins, Sixth New York Independent Battery, 1861–1864*. Knoxville: University of Tennessee Press, 2006.

Gross, S. W. "On Torpedo Wounds." *American Journal of Medical Sciences* 51 (April 1866): 369–72.

Hagemann, E. R., ed. *Fighting Rebels and Redskins: Experiences in Army Life of Colonel George B. Sanford, 1861–1892*. Norman: University of Oklahoma Press, 1969.

Hall, Henry, and James Hall. *Cayuga in the Field: A Record of the 19th N Y Volunteers, All the Batteries of the 3rd New York Artillery, and 75th New York Volunteers*. Syracuse: Truair & Smith, 1873.

Hall, Hillman A. *History of the Sixth New York Cavalry (Second Ira Harris Guard)*. Worcester, Massachusetts: Blanchard Press, 1908.

Halleck, H. Wager. *Elements of Military Art and Science: Or, Course of Instruction in Strategy, Fortification, Tactics of Battles*. New York: D. Appleton, 1862.

Harrison, Burton. "The Capture of Jefferson Davis." *Century Magazine* 27, no. 1 (November 1883): 130–45.

Harwell, Richard Barksdale, and Philip N. Racine, eds. *The Fiery Trail: A Union Officer's Account of Sherman's Last Campaigns*. Knoxville: University of Tennessee Press, 1986.

Hatch, Carl E., ed. *Dearest Susie: A Civil War Infantryman's Letters to His Sweetheart*. New York: Exposition Press, 1971.

Haupt, Herman. *Military Bridges: With Suggestions of New Expedients and Constructions for Crossing Streams and Chasms*. New York: D. Van Nostrand, 1864.

Haupt, Herman. *Reminiscences*. Milwaukee, Wisconsin: Wright and Joys, 1901.

Hays, Gilbert Adams, comp. *Under the Red Patch: Story of the Sixty-Third Regiment Pennsylvania Volunteers, 1861–1864*. Pittsburgh: Market Review, 1908.

Hazen, William B. *A Narrative of Military Service*. Boston: Ticknor, 1885.

Hearn, Chester G. *Mobile Bay and the Mobile Campaign: The Last Great Battles of the Civil War*. Jefferson, North Carolina: McFarland, 1993.

Hedley, F. Y. *Marching Through Georgia*. Chicago: Donohue, Henneberry, 1890.

Heitman, Francis H. *Historical Register and Dictionary of the United States Army*. 2 vols. Washington, DC: Government Printing Office, 1903.

Hess, Earl J. *Civil War Logistics: A Study of Military Transportation*. Baton Rouge: Louisiana State University Press, 2017.

Hess, Earl J. *The Civil War in the West: Victory and Defeat from the Appalachians to the Mississippi*. Chapel Hill: University of North Carolina Press, 2012.

Hess, Earl J. *Field Armies and Fortifications in the Civil War: The Eastern Campaigns, 1861–1864*. Chapel Hill: University of North Carolina Press, 2005.

Hess, Earl J. *In the Trenches at Petersburg: Field Fortifications & Confederate Defeat*. Chapel Hill: University of North Carolina Press, 2009.

Hills, Charles S. "The Last Battle of the War: Recollections of the Mobile Campaign." In *War Papers and Personal Reminiscences, 1861–1865: Read Before the Commandery of the State of Missouri, Military Order of the Loyal Legion of the United States*. Vol. 1. St. Louis: Becktold, 1892: 177–290.

History of the Fifth Massachusetts Battery. Boston: Luther E. Cowles, 1902.

History of the Forty-Sixth Regiment Indiana Volunteer Infantry, September, 1861– September, 1865. Logansport: Wilson and Humphreys, 1888.

Hogane, J. T. "Reminiscences of the Siege of Vicksburg, Pt. 2." *Southern Historical Society Papers* 11 (1883): 282–97.

Holbrook, William C. *A Narrative of the Services of the Officers and Enlisted Men of the 7th Regiment of Vermont Volunteers (Veterans), from 1862 to 1866*. New York: American Bank Note Company, 1882.

Hoole, William Stanley. *Alabama Tories: The First Alabama Cavalry, U.S.A., 1862– 1865*. Tuscaloosa, Alabama: Confederate Publishing, 1960.

Howard, Charles H. "Incidents and Operations Connected with the Capture of Savannah." In *Military Essays and Recollections: Papers Read Before the Commandery of the State of Illinois, Military Order of the Loyal Legion of the United States*. Vol. 4. Chicago: Cozzens and Beaton, 1907: 430–50.

Howard, Oliver Otis. *Autobiography*. 2 vols. New York: Baker and Taylor, 1907.

Howard, R. L. *History of the 124th Regiment Illinois Infantry Volunteers*. Springfield, Illinois: H. W. Rokker, 1880.

Howe, M. A. DeWolfe, ed. *Marching with Sherman: Passages from the Letters and Campaign Diaries of Henry Hitchcock*. Lincoln: University of Nebraska Press, 1995.

Huffine, L. C. "Torpedoes at Fort McAllister." *National Tribune*, March 14, 1907.

Hughes, Nathaniel Cheairs, Jr., ed. *The Civil War Memoir of Philip Daingerfield Stephenson, D.D.* Conway: University of Central Arkansas Press, 1995.

Hughes, Nathaniel Cheairs, Jr., ed. *Liddell's Record: St. John Richardson Liddell, Brigadier General, CSA*. Dayton, Ohio: Morningside, 1985.

Hunter, Antwain K. "'Patriots,' 'Cowards,' and 'Men Disloyal at Heart': Labor and Politics at the Springfield Armory, 1861–1865," *Journal of Military History* 84, no. 1 (January 2020): 51–81.

Hutcheon, Wallace, Jr. *Robert Fulton: Pioneer of Undersea Warfare*. Annapolis, Maryland: Naval Institute Press, 1981.

Hyde, Thomas W. *Following the Greek Cross; Or, Memories of the Sixth Army Corps*. Columbia: University of South Carolina Press, 2005.

Hyde, William L. *History of the One Hundred and Twelfth Regiment, N.Y. Volunteers*. Fredonia, New York: W. McKinstry, 1866.

Index of Patents Relating to Electricity, Granted by the United States Prior to July 1, 1881, with an Appendix Embracing Patents Granted from July 1, 1881, to June 30, 1882. Washington, DC: Government Printing Office, 1882.

"Infernal Machines in the Mississippi." *Scientific American* 6, no. 14 (April 5, 1862): 210–11.

"The Invention of Torpedoes: Gen. Gabriel J. Rains, of South Carolina, Bears the Honor." *Confederate Veteran* 2 (1894): 234–35.

Jackson, Harry L. *First Regiment Engineer Troops, P.A.C.S.: Robert E. Lee's Combat Engineers*. Louisa, Virginia: R. A. E. Design and Publishing, 1998.

Jackson, Joseph Orville, ed. *"Some of the Boys": The Civil War Letters of Isaac Jackson, 1862–1865*. Carbondale: Southern Illinois University Press, 1960.

Jackson, Oscar L. *The Colonel's Diary*. N.p., 1922.

James, David H. *The Siege of Port Arthur: Records of an Eye-Witness*. London: T. Fisher Unwin, 1905.

Johnson, Charles Beneulyn. *Muskets and Medicine: Or Army Life in the Sixties*. Philadelphia: F. A. Davis, 1917.

Johnson, William L. "The Assault of Fort McAllister." *National Tribune*, July 19, 1883.

Johnston, Joseph E. "Manassas to Seven Pines." Vol. 2 of *Battles and Leaders of the Civil War*, edited by Robert Underwood Johnson and Clarence Clough Buel. New York: Thomas Yoseloff, 1956: 202–18.

Jones, Charles Colcock, Jr. *Historical Sketch of the Chatham Artillery, During the Confederate Struggle for Independence*. Albany, New York: Joel Munsell, 1867.

Jones, Charles Colcock, Jr. *The Siege of Savannah in December, 1864, and the Confederate Operations in Georgia*. Albany, New York: Joel Munsell, 1874.

Jones, Charles H. *Artillery Fuses of the Civil War*. Alexandria, Virginia: O'Donnell Publications, 2001.

Jones, Ian. *Malice Aforethought: The History of Booby Traps from World War One to Vietnam*. Mechanicsburg, Pennsylvania: Stackpole Books, 2004.

Joyce, Fred. "Infantry Stampede." *Southern Bivouac* 2, no. 5 (January, 1884): 223–25.

Kerner, Robert J., ed. "The Diary of Edward W. Crippen, Private 27th Illinois Volunteers, War of the Rebellion, August 7, 1862, to September 19, 1864." *Publications of the Illinois State Historical Society*, no. 14 (1911): 220–82.

King, W. R. *Torpedoes: Their Invention and Use, from the First Application to the Art of War to the Present Time, for the Use of the Officers of the Corps of Engineers*. Washington, DC: Government Printing Office, 1866.

Kirwan, A. D., ed. *Johnny Green of the Orphan Brigade: The Journal of a Confederate Soldier*. Lexington: University Press of Kentucky, 2002.

Kochan, Michael P., and John C. Wideman. *Civil War Torpedoes: A History of Improvised Explosive Devices in the War Between the States*. DVD, 2012.

Lamb, William. "Fort Fisher." *Southern Historical Society Papers* 21 (1893): 257–90.

Lamb, William. "Thirty-Sixth Regiment (Second Artillery)." In *Histories of the Several Regiments and Battalions from North Carolina in the Great War 1861–1865*. Vol. 2. Goldsboro, North Carolina: Nash Brothers, 1901: 629–51.

Lasswell, Mary E., ed. *Rags and Hope: The Recollections of Val C. Giles, Four Years with Hood's Brigade, Fourth Texas Infantry, 1861–1865*. New York: Coward-McCann, 1961.

Laumer, Frank, ed. *Amidst a Storm of Bullets: The Diary of Lt. Henry Prince in Florida, 1836–1842*. Tampa, Florida: University of Tampa Press, 1998.

"Letter from a Sharpshooter," May 4, 1862, *Rutland Weekly Herald*, May 15, 1862.

Lewis, S. Joseph, Jr., ed. "Letters of William Fisher Plane, C.S.A. to His Wife." *Georgia Historical Quarterly* 48, no. 1 (March 1964): 215–28.

Lieber, Francis. *Contributions to Political Science, Including Lectures on the Constitution of the United States and Other Papers . . . Being Volume II of His Miscellaneous Writings*. Philadelphia: J. B. Lippincott, 1881.

Lieber, Francis. *Reminiscences, Addresses, and Essays . . . Being Volume I of His Miscellaneous Writings*. Philadelphia: J. B. Lippincott, 1881.

Little, George, and James R. Maxwell. *A History of Lumsden's Battery, C.S.A.* Tuscaloosa, AL: R. E. Rhodes Chapter United Daughters of the Confederacy, [1905].

Livingston, Gary. *"Among the Best Men the South Could Boast": The Fall of Fort McAllister, December 13, 1864*. Cooperstown, NY: Caisson Press, 1997.

Lockwood, Thomas D. *Electricity, Magnetism, and Electric Telegraphy: A Practical Guide and Hand-Book*. New York: D. Van Nostrand, 1883.

Lomax, W. W. "The Torpedoes at Savannah." *National Tribune*, December 26, 1901.

Lord, Francis A. "Both Sides Used Torpedoes Widely." *Civil War Times Illustrated* 2, no. 9 (January 1964): 46–48.

Lord, Francis A. *Civil War Collector's Encyclopedia: Arms, Uniforms, and Equipment of the Union and Confederacy*. Harrisburg, Pennsylvania: Stackpole, 1963.

Lossing, Benson J. *Pictorial Field Book of the Civil War*. 3 vols. Baltimore: Johns Hopkins University Press, 1997.

Lucas, D. R. *New History of the 99th Indiana Infantry*. Rockford, Illinois: Horner Printing, 1900.

Lundeberg, Philip K. *Samuel Colt's Submarine Battery: The Secret and the Enigma*. Washington, DC: Smithsonian Institution Press, 1974.

Lyon, Adelia C., comp. *Reminiscences of the Civil War*. San Jose, California: Muirson & Wright, 1907.

Mahon, John K. *History of the Second Seminole War, 1835–1842*. 2nd ed. Gainesville: University Press of Florida, 1985.

Malles, Ed, ed. *Bridge Building in Wartime: Colonel Wesley Brainerd's Memoirs of the 50th New York Volunteer Engineers*. Knoxville: University of Tennessee Press, 1997.

Malone, Dumas, ed. *Dictionary of American Biography*. 20 vols. New York: Charles Scribner's Sons, 1928–1937.

Manual of Field Works (All Arms). London: His Majesty's Stationery Officer, 1921.

March, L. B. "Hunting Torpedoes Near Savannah." *National Tribune*, August 14, 1902.

Marks, J. J. *The Peninsular Campaign in Virginia, or Incidents and Scenes on the Battle-fields and In Richmond*. Philadelphia: J. B. Lippincott, 1864.

Marshall, Albert O. *Army Life: From a Soldier's Journal*. Fayetteville: University of Arkansas Press, 2009.

Marshall, T. B. *History of the Eighty-Third Ohio Volunteer Infantry: The Greyhound Regiment*. Cincinnati, Ohio: Gibson and Perin, 1913.

Maury, Dabney H. "Defence of Spanish Fort: Some Comment by the Confederate Commander on Mr. P. D. Stephenson's Article." *Southern Historical Society Papers* 39, no. 1 (April 1914): 130–36.

Medical and Surgical History of the Civil War. 12 vols. Wilmington, NC: Broadfoot Publishing, 1991.

Messent, Peter, and Steve Courtney, eds. *The Civil War Letters of Joseph Hopkins Twichell: A Chaplain's Story*. Athens: University of Georgia Press, 2006.

Miers, Earl Schenck, ed. *A Rebel War Clerk's Diary*. New York: Sagamore Press, 1958.

Millett, John W. "At Port Hudson: A Boy of the 24th Me. Has Some Exciting Adventures." *National Tribune*, August 19, 1909.

Minnich, J. W. "Incidents of the Peninsular Campaign." *Confederate Veteran* 30 (1922): 53–56.

Moore, Frank. *The Rebellion Record*. 10 vols. New York: D. Van Nostrand, 1861–1867.

Mulligan, James A. "The Siege of Lexington, MO." In *Battles and Leaders of the Civil War*. Vol. 1, edited by Robert Underwood Johnson and Clarence Clough Buel. New York: Thomas Yoseloff, 1956: 307–13.

Nanzig, Thomas P., ed. *The Civil War Memoirs of a Virginia Cavalryman: Lt. Robert T. Hubard, Jr.* Tuscaloosa: University of Alabama Press, 2007.

National Cyclopaedia of American Biography. 63 vols. Ann Arbor, Michigan: University Microfilms, 1967.

Nevins, Allan, ed. *A Diary of Battle: The Personal Journals of Colonel Charles S. Wainwright, 1861–1865*. New York: Harcourt, Brace, & World, 1962.

Nichols, George Ward. *The Story of the Great March*. New York: Harper and Brothers, 1866.

Nichols, James L. *Confederate Engineers*. Tuscaloosa, Alabama: Confederate Publishing, 1957.

Ninth Reunion of the 37th Regiment O.V.V.I. Toledo, Ohio: Montgomery & Vrooman, 1890.

Nowlin, S. H. "Capture and Escape of S. H. Nowlin, Private Fifth Virginia Cavalry." *Southern Bivouac* 2 (October 1883): 70–73.

O'Brien, Sean Michael. *Mobile, 1865: Last Stand of the Confederacy*. Westport, Connecticut: Praeger, 2001.

O'Brien. W. D. "They Planted Shells." *National Tribune*, June 17, 1886.

The Official Military Atlas of the Civil War. New York: Fairfax Press, 1983.

Official Records of the Union and Confederate Navies in the War of the Rebellion. 30 vols. Washington, DC: Government Printing Office, 1894–1922.

Official Roster of the Soldiers of the State of Ohio in the War of the Rebellion, 1861–1865. 12 vols. Akron: Werner, 1886–1895.

"Operations in the Interior: Additional Particulars of the Movement." *New York Times*, February 17, 1865.

Oxford English Dictionary. 2nd ed. 20 vols. Oxford, United Kingdom: Clarendon Press, 1989.

Pack, Reynell. *Sebastopol Trenches and Five Months in Them*. London: Kerby and Endean, 1878.

Palfrey, J. C. "The Capture of Mobile, 1865." *The Mississippi Valley, Tennessee, Georgia, Alabama, 1861–1864, Vol. 8: Papers of the Military Historical Society of Massachusetts*. Boston: Military Historical Society of Massachusetts, 1910: 529–57.

Palladino, Anita, ed. *Diary of a Yankee Engineer: The Civil War Story of John H. Westervelt, Engineer, 1st New York Volunteer Engineer Corps*. New York: Fordham University Press, 1997.

Parker, John L. *History of the Twenty-Second Massachusetts Infantry*. Boston: Rand Avery, 1887.

Parker, Thomas J. "The Capture of Fort Esperanza, Texas." *National Tribune*, August 16, 1883.

Patrick, Jeffrey L., and Robert J. Willey, eds. *Fighting for Liberty and Right: The Civil War Diary of William Bluffton Miller, First Sergeant, Company K, Seventy-Fifth Indiana Volunteer Infantry*. Knoxville: University of Tennessee Press, 2005.

Perry, Milton F. *Infernal Machines: The Story of Confederate Submarine and Mine Warfare*. Baton Rouge: Louisiana State University Press, 1965.

Pierce, W. S. "In Command of a Springfield." *National Tribune*, August 25, 1910.

Pleydell, J. C. *An Essay on Field Fortification Intended Principally for the Use of Officers of Infantry*. London: F. Wingrave, 1768.

Plum, William R. *The Military Telegraph during the Civil War in the United States*. 2 vols. Chicago: Jansen, McClurg, 1882.

Polk, William M. *Leonidas Polk: Bishop and General*. 2 vols. London: Longmans, Green, 1893.

Popchock, Barry., ed. *Soldier Boy: The Civil War Letters of Charles O. Musser, 29th Iowa*. Iowa City: University of Iowa Press, 1995.

Porter, Charles. "Explosion at City Point." *National Tribune*, January 7, 1904.

Porter, David Dixon. "Torpedo Warfare." *North American Review* 127 (September–October 1878): 213–36.

Prescott, George B. *History, Theory, and Practice of the Electric Telegraph*. Boston: Ticknor and Fields, 1860.

Price, Isaiah. *History of the Ninety-Seventh Regiment, Pennsylvania Volunteer Infantry, during the War of the Rebellion, 1861–65*. Philadelphia: B & P Printers, 1875.

Private and Official Correspondence of Gen. Benjamin F. Butler: During the Period of the Civil War. 5 vols. Norwood, Massachusetts: Plimpton Press, 1917.

Pyne, Henry R. *The History of the First New Jersey Cavalry*. Trenton, New Jersey: J. A. Beecher, 1871.

Ragan, Mark K. *Confederate Saboteurs: Building the Hunley and Other Secret Weapons of the Civil War*. College Station: Texas A&M University Press, 2015.

Rains, G. J. "Torpedoes." *Southern Historical Society Papers* 3 (1887): 255–60.

Ramsey, Albert C., trans. *The Other Side: Or Notes for the History of the War between Mexico and the United States*. New York: John Wiley, 1850.

Ratchford, J. W. "More of Gen. Rains and His Torpedoes." *Confederate Veteran* 2 (1894): 283.

Read, George W. "Gen. Osterhaus Was Cool." *National Tribune*, September 15, 1904.

Reichardt, Theodore. *Diary of Battery A, First Regiment Rhode Island Light Artillery*. Providence: N. Bangs Williams, 1865.

Rhea, Gordon C. *The Battles for Spotsylvania Court House and the Road to Yellow Tavern, May 7–12, 1864*. Baton Rouge: Louisiana State University Press, 1997.

Rhodes, John H. *The History of Battery B, First Regiment Rhode Island Light Artillery in the War to Preserve the Union, 1861–1865*. Providence: Snow & Farnham, 1894.

Richter, Donald. *Chemical Soldiers: British Gas Warfare in World War I*. Lawrence: University Press of Kansas, 1992.

Robins, Colin, ed. *Captain Dunscombe's Diary: The Real Crimean War that the British Infantry Knew*. Bowdon, United Kingdom: Withycut House, 2003.

Robson, George Jr. "Star-Spangled Land Mines." *Military Engineer* 49 (September–October 1957): 354.

Roche, Roberta Senechal de la, ed. *"Our Aim Was Man": Andrew's Sharpshooters in the American Civil War*. Amherst: University of Massachusetts Press, 2016.

Rottman, Gordon L. *World War II Axis Booby Traps and Sabotage Tactics*. Midland House, West Way, Botley, Oxford, England: Osprey, 2009.

Rowland, Dunbar, ed. *Jefferson Davis, Constitutionalist: His Letters, Papers and Speeches*. 10 vols. New York: J. J. Little & Ives, 1923.

Rutherford, Kenneth R. *America's Buried History: Landmines in the Civil War*. El Dorado Hills, California: Savas Beatie, 2020.

Salecker, Gene Eric. *Disaster on the Mississippi: The Sultana Explosion, April 27, 1865*. Annapolis, Maryland: Naval Institute Press, 1996.

Sater, William F. *Andean Tragedy: Fighting the War of the Pacific, 1879–1884*. Lincoln: University of Nebraska Press, 2007.

Saunier, Joseph A., ed. *A History of the Forty-Seventh Regiment Ohio Veteran Volunteer Infantry*. Hillsboro, Ohio: Lyle Printing, [1903?].

Schafer, Louis S. *Confederate Underwater Warfare: An Illustrated History*. Jefferson, North Carolina: McFarland, 1996.

Schaff, Morris. "The Explosion at City Point." In *Civil War Papers Read Before the Commandery of the State of Massachusetts, Military Order of the Loyal Legion of the United States*. Vol. 2. Wilmington, North Carolina: Broadfoot Publishing, 1993: 477–85.

Schellen, H. *Magneto-Electric and Dynamo-Electric Machines: Their Construction and Practical Application to Electric Lighting and the Transmission of Power*. New York: D. Van Nostrand, 1884.

Schiffer, Michael Brian. *Power Struggles: Scientific Authority and the Creation of Practical Electricity before Edison*. Cambridge, Massachusetts: MIT Press, 2008.

Schiller, Herbert M., ed. *Confederate Torpedoes: Two Illustrated 19th Century Works with New Appendices and Photographs*. Jefferson, North Carolina: McFarland, 2011.

Schneck, William C. "The Origins of Military Mines: Part I." *Engineer* 28 (July 1998): 49–55.

Schneck, William C. "The Origins of Military Mines: Part II." *Engineer* 28 (November 1998): 44–50.

Scott, H. L. *Military Dictionary: Comprising Technical Definitions, Information on Raising and Keeping Troops, Actual Service, Including Makeshifts and Improved Matériel, and Law, Government, Regulation, and Administration Relating to Land Forces.* New York: D. Van Nostrand, 1861.

Scott, John, comp. *Story of the Thirty-Second Iowa Infantry Volunteers.* Nevada, Iowa: John Scott, 1896.

Scott, R. B. *The History of the 67th Regiment Indiana Infantry Volunteers, War of the Rebellion.* Bedford, Indiana: Herald, 1892.

Sears, Stephen W., ed. *The Civil War Papers of George B. McClellan: Selected Correspondence, 1860–1865.* New York: Ticknor and Fields, 1989.

Sears, Stephen W. *To the Gates of Richmond: The Peninsula Campaign.* New York: Ticknor & Fields, 1992.

Shannon, I. N. "'Infernal Machines' Described." *Confederate Veteran* 13 (1905): 458.

Sharland, George. *Knapsack Notes of Sherman's Campaign Through the State of Georgia.* Springfield, Illinois: Johnson and Bradford, 1865.

Shepley, George F. "Incidents of the Capture of Richmond." *Atlantic Monthly* 46 (1880): 18–28.

Sheridan, P. H. *Personal Memoirs.* 2 vols. Wilmington, North Carolina: Broadfoot Publishing, 1992.

Sherman, William T. *Memoirs.* 2 vols. New York: D. Appleton, 1875.

Sigal, Leon V. *Negotiating Minefields: The Landmines Ban in American Politics.* New York: Routledge, 2006.

Simmons, William. "Confederate Torpedoes." *National Tribune*, April 28, 1892.

Simon, John Y., ed. *The Papers of Ulysses S. Grant.* 28 vols. Carbondale: Southern Illinois University Press, 1967–2005.

Slocum, Charles Elihu. *The Life and Services of Major-General Henry Warner Slocum.* Toledo, Ohio: Slocum Publishing, 1913.

Smith, Cyrus E. "Capturing Fort Blakely." *National Tribune*, February 3, 1910.

Smith, H. I. *History of the Seventh Iowa Veteran Volunteer Infantry During the Civil War.* Mason City, Iowa: E. Hitchcock, 1903.

Smith, Justin H. *The War with Mexico.* 2 vols. New York: Macmillan, 1919.

Snyder, Dean. "Torpedoes for the Confederacy." *Civil War Times Illustrated* 24, no. 1 (March 1985): 40–45.

Sommers, Richard J. *Richmond Redeemed: The Siege at Petersburg.* Garden City, New York: Doubleday, 1981.

Sorrel, G. Moxley. *Recollections of a Confederate Staff Officer.* New York: Neale, 1905.

Sperry, A. F. *History of the 33rd Iowa Infantry Volunteer Regiment, 1863–6.* Fayetteville: University of Arkansas Press, 1999.

Stephenson, P. D. "Defence of Spanish Fort: On Mobile Bay—Last Great Battle of the War." *Southern Historical Society Papers* 39, no. 1 (April, 1914): 118–29.

Still, William N., Jr., ed. "The Civil War Letters of Robert Tarleton." *Alabama Historical Quarterly* 32, no. 1 & 2 (Spring and Summer 1970): 52–80.

Stimson, J. B. "Three Unusual Gun-Shot Wounds." *The Southern Practitioner* 22, no. 8 (August 1900): 363–64.

Stone, DeWitt Boyd, Jr., ed. *Wandering to Glory: Confederate Veterans Remember Evans' Brigade*. Columbia: University of South Carolina Press, 2002.

Stoneburner, E. A. "The Siege of Blakely: Comrade Stoneburner Writes of His Experiences With the 114th Ohio." *National Tribune*, November 30, 1899.

Stormont, Gilbert R., comp. *History of the Fifty-Eighth Regiment of Indiana Volunteer Infantry*. Princeton, Indiana: Clarion Press, 1895.

Story of the Fifty-Fifth Regiment Illinois Volunteer Infantry in the Civil War, 1861–1865. Clinton, Massachusetts: W. J. Coulter, 1887.

Strong, William E. "Capture of Fort McAllister." *Georgia Historical Quarterly* 88, no. 3 (2004): 406–21.

Sumner, Merlin E., ed. *The Diary of Cyrus B. Comstock*. Dayton, Ohio: Morningside Bookshop, 1987.

Supplement to the Official Records of the Union and Confederate Armies. 100 vols. Wilmington, North Carolina: Broadfoot Publishing, 1995–1999.

Tarrant, E. W. "Siege and Capture of Fort Blakely." *Confederate Veteran* 23 (1915): 457–58.

Taylor, Gary Nelson, ed. *Saddle and Saber: The Letters of Civil War Cavalryman Corporal Nelson Taylor*. Bowie, Maryland: Heritage Books, 1993.

Terry, Carlisle. "An Anecdote of Jefferson Davis." *Century* 39, issue 4 (February 1890): 638

Throne, Mildred, ed. "A History of Company D, Eleventh Iowa Infantry, 1861–1865." *Iowa Journal of History* 55, no. 1 (January 1957): 35–90.

Tidwell, William A., James O. Hall, and David Winfred Gaddy. *Come Retribution: The Confederate Secret Service and the Assassination of Lincoln*. Jackson: University Press of Mississippi, 1988.

Tielke, J. G. *The Field Engineer: Or Instructions Upon Every Branch of Field Fortification*. 2 vols. London: J. Walter, 1789.

Todd, Glenda McWhirter. *First Alabama Cavalry, U.S.A.: Homage to Patriotism*. Bowie, Maryland: Heritage Books, 1999.

Toepfer. "Technics in the Russo-Japanese War." *Professional Memoirs, Corps of Engineers, United States Army, and Engineer Department at Large* 2, no. 6 (April–June 1910): 174–201.

Tourgée, Albion W. *The Story of a Thousand: Being a History of the Service of the 105th Ohio Volunteer Infantry, in the War for the Union from August 21, 1862 to June 6, 1865*. Buffalo, New York: S. McGerald & Son, 1896.

Trautmann, Frederic, ed. *A Prussian Observes the American Civil War: The Military Studies of Justus Scheibert*. Columbia: University of Missouri Press, 2001.

Trounce, H. D. *Fighting the Boche Underground*. New York: Charles Scribner's Sons, 1918.

Trout, Robert J., ed. *Memoirs of the Stuart Horse Artillery Battalion: Moorman's and Hart's Batteries*. Knoxville: University of Tennessee Press, 2008.

Trowbridge, John T. *The Desolate South, 1865–1866*. Freeport, New York: Books for Libraries, 1970.

Trusty, Lance. "Private Smith Takes Mobile: A Soldier's Journal." *Lincoln Herald* 80, no. 2 (Summer 1978): 78–83.

Vance, J. W., ed. *Report of the Adjutant General of the State of Illinois.* 8 vols. Springfield, Illinois: H. W. Rokker, 1886.

Vaughan, Taylor. "A Johnny's View: Gives a Few Points about Blakely as He Knew Them." *National Tribune*, February 22, 1900.

Wagner, Phil M. "When at Spanish Fort: Amusing Reminiscences of the Time Just Before Evacuation." *National Tribune*, February 24, 1898.

Walke, H. *Naval Scenes and Reminiscences of the Civil War in the United States on the Southern and Western Waters.* New York: F. R. Reed, 1877.

Walton, Clyde C., ed. *Behind the Guns: The History of Battery I, 2nd Regiment, Illinois Light Artillery.* Carbondale; Southern Illinois University Press, 1965.

The War of the Rebellion: A Compilation of the Official Records of the Union and Confederate Armies. 70 vols. in 128. Washington, DC: Government Printing Office, 1880–1901.

Waters, W. Davis. "'Deception Is the Art of War': Gabriel J. Rains, Torpedo Specialist of the Confederacy." *North Carolina Historical Review* 66, no. 1 (January 1989): 29–60.

Waters, W. Davis, and Joseph I. Brown. *Gabriel Rains and the Confederate Torpedo Bureau.* Durham, North Carolina: Monograph, 2014.

Way, Virgil G., comp. *History of the Thirty-Third Regiment Illinois Veteran Volunteer Infantry.* Gibson City, Ilinois: Gibson Courier, 1902.

Whiteshot, Charles Austin. *The Oil-Well Driller: A History of the World's Greatest Enterprise, the Oil Industry.* Mannington, West Virginia: Charles Austin Whiteshot, 1905.

Wideman, John C. *The Sinking of the USS Cairo.* Jackson: University Press of Mississippi, 1993.

Williams, Jody, Stephen D. Goose, and Mary Wareham, eds. *Banning Landmines: Disarmament, Citizen Diplomacy, and Human Security.* Lanham, Maryland: Rowman and Littlefield, 2008.

Williams, T. Harry, ed. *With Beauregard in Mexico: The Mexican War Reminiscences of G. T. Beauregard.* Baton Rouge: Louisiana State University Press, 1956.

Wilson, James H. "The Cavalry of the Army of the Potomac." *Papers of the Military Historical Society of Massachusetts*. Vol. 13. Wilmington, North Carolina: Broadfoot, 1990: 35–88.

Wilson, John M. "The Campaign Ending with the Capture of Mobile." *Military Order of the Loyal Legion of the United States, Commandery of the District of Columbia, War Papers No. 17.* N.p., n.d.

Winschel, Terrence J., ed. *The Civil War Diary of a Common Soldier: William Wiley of the 77th Illinois Infantry.* Baton Rouge: Louisiana State University Press, 2001.

Winter, William C., ed. *Captain Joseph Boyce and the 1st Missouri Infantry, C.S.A.* St. Louis: Missouri History Museum, 2011.

Wise, Stephen R. *Gate of Hell: Campaign for Charleston Harbor, 1863.* Columbia: University of South Carolina Press, 1994.

Witt, John Fabian. *Lincoln's Code: The Laws of War in American History*. New York: Simon and Schuster, 2012.

Wittenberg Eric J., ed. *"We Have It Damn Hard Out Here": The Civil War Letters of Sergeant Thomas W. Smith, 6th Pennsylvania Cavalry*. Kent, Ohio: Kent State University Press, 1999.

Wolters, Timothy S. "Electric Torpedoes in the Confederacy: Reconciling Conflicting Histories." *Journal of Military History* 72, no. 3 (July 2008): 755–83.

Wood, Wales W. *A History of the Ninety-Fifth Regiment Illinois Infantry Volunteers, from Its Organization in the Fall of 1862, until Its Final Discharge from the United States Service, in 1865*. Chicago: Tribune, 1865.

Woods, J. T. *Services of the Ninety-Sixth Ohio Volunteers*. Toledo, Ohio: Blade, 1874.

Wright, T. J. *History of the Eighth Regiment Kentucky Vol. Inf.* St. Joseph, Missouri: St. Joseph Steam Printing, 1880.

Yeary, Mamie, comp. *Reminiscences of the Boys in Gray, 1861–1865*. Dallas, Texas: Smith-Lamar, 1912.

Youngblood, Norman. *The Development of Mine Warfare: A Most Murderous and Barbarous Conduct*. Westport, Connecticut: Praeger, 2006.

Index

Abel, Frederick Augustus, 102, 103, 158, 171
Alabama units, CS: Fourth Cavalry, 142; Forty-first Infantry, 78, 196
Alabama units, US: First Cavalry, 83, 97
Alexander, Edward Porter, 31, 74, 168
The Alleghanian, 30
Allen, Charles J., 114, 126, 127, 173
American Revolution, 2
Ammen, Daniel, 15
Anderson, George W., 89, 93, 94
Andrews, Christopher C., 120, 123
Andrews, Welburn J., 48
Army of Tennessee, 69
Atlanta campaign, 140, 143, 144, 162
Atlanta History Center, 42

Babcock, Orville E., 150
Badajoz, siege of, 2
Baird, Absalom, 95
Baker, J. T., 63, 64
Baker, L. C., 156
Baldwin, James S., 55, 59
Barnes, Amos P., 133
Barnes, John S., xviii, 148, 149, 171
Barry, William F., 42, 188, 189
Batchelder's Creek, North Carolina, xv, 131–33, 193
Battery Wagner, South Carolina, 35

Beacher, Henry S., 123
Beardslee, George W., 156, 157, 159, 160
Beauregard, P. G. T., 2, 51–53, 59, 61, 82, 83, 110, 140, 162, 186
Beggs, Thomas D., 20
Belknap, William W., 165
Bennett, James Gordon, 11
Bir, Louis, 115
Birney, William, 138
Blair, Frank P., 83, 84, 87
Blake, Peleg W., 24
Blakely, Alabama, 119–29, 173, 194
Bombaugh, Charles E., 24, 26, 28
Boteler, Alexander R., 142
Bower, Adam, 184
Bradbury, David, 63–65, 145
Bradfute, William R., 63
Bremfoerder, Henry, 90
Brooks, Thomas B., 54–57, 59, 61, 62
Browne, William M., 82
Buckman, George R., 29, 30
Burke, John W., 169
Burkhalter, Jerry, 78, 196
Bushnell, Wells A., 71
Butler, Benjamin, 100, 133, 148
Buzzell, Andrew J. H., 105
Byrnes, William, 30

Campbell, John A., 140
Canby, Edward R. S., 114, 117, 126, 128
Canstadt, Baron Pavel Schilling von, 6
Carolinas campaign, 106–11, 169
Carrington, George, 115
Century Magazine, 169
Chace, Charles, 65
Chancellorsville campaign, 135
Chapultepec, battle of, 2, 3
Chattanooga Gazette, 146
Chichester, C. E., 59
Cincinnati Gazette, 118
City Point, Virginia, xv, 149–53, 158, 170, 193
Ciudad Rodrigo, siege of, 2
Civil War: global landmine warfare, xiii–xv, 159, 166, 175, 177, 185, 190, 192; number of landmine casualties in, xv, 193
Claassen, Peter J., 132
Clayton, William H. H., 128
Cleveland, Moses A., 115, 119, 124, 125
Cobb, Howell, 82
Cold War, xiv, 181–84, 190, 191
Colt, Samuel, 6
Columbus, 148
Columbus, Kentucky, 13, 15–20, 155, 192
Comstock, Cyrus B., 100–102, 104, 158
Confederate Veteran, 167
Connecticut units: Second Light Battery, 122
Conscription Bureau, 39, 41
Convention on Prohibition or Restrictions on the Use of Certain Conventional Weapons, 182
Convention on the Prohibition of the Use, Stockpiling, Production and Transfer of Anti-Personnel Mines and on their Destruction (Antipersonnel Mine Ban Treaty), 183, 184
Cooper, Samuel, 39
Courtenay, Thomas E., 147, 169

Crawford, Lewis, 150
Crimean War, xiii, 6–10, 38, 53, 127, 155, 175, 177, 192
Crippen, Edward W., 19
Croll, Mike, xvi, 3, 182, 186
Cross, J. L., 43
Crowell, William C., 14
CSS Albemarle, 131, 133
CSS Chicora, 82
Cullum, George W., 17, 19
Cusick, Charles C., 132

Dabney, Frederick T., 45
Dahlgren, John A., 93
Dale (or Deal), Charlie, 149
Danish War, 157
Davidson, Hunter, 167
Davis, Jefferson, 4, 39, 43, 44, 46, 48, 67–69, 72, 74, 81, 82, 110, 111, 145, 147, 148, 162, 166–69, 187, 188
Davis, Jefferson C., 87, 108
Delafield, Richard, 8–10, 127, 157, 170
Deupree, William S., 78
Dillard, R. K., 149, 150, 152
Dodge, Grenville M., 142
Doubleday, Ulysses, 139
Drish, James F., 117

Eddington, William R., 121–23
Edwards, George, 138
Efaw, Andrew C. S., 183
Elder, William D., 68
Ernst, Oswald H., 165
Evans, Thomas L., 126
Ewing, Hugh Boyd, 47

Faraday, Michael, 5
Fawcett, William, 135, 137
Ferguson, John Hill, 109
Fetterman, George W., 118
Fike, Henry, 118
Finn, W. H., 195
First Persian Gulf War, 182
Floyd, Fred C., 30
Fontaine, Lamar, 142

Foote, Andrew H., 16, 17
Fordice, John R., 196
Forrest, Nathan Bedford, 142
Fort Beauregard, Port Royal, South Carolina, 14, 15
Fort Esperanza, Texas, 62–65, 67
Fort Fisher, North Carolina, 99–106, 155, 158, 170, 192
Fort Griffin, Texas, 65
Fort King, Florida, 4, 5, 19, 33, 36, 155, 166, 167, 187, 189–91
Fort McAllister, Georgia, 89–94, 140, 161, 166, 193, 194, 208nn26–28, 209n37
Fort Wagner, South Carolina, 51–62, 67, 194, 204n31
Foster, John G., 108, 137
Fougasse, xiii, 1, 2, 45, 63–65, 170, 171
Fox, George Benson, 56
Fox, Gustavus Vasa, 149
Frank Leslie's Illustrated Newspaper, 17, 20
Fredericksburg campaign, 135, 160
Fretwell, John R., 74, 145

Gallipoli campaign, 177
Galvani, Luigi, 5
galvanic battery, 5, 6, 18, 155, 156, 159
Gates, Elijah, 124
Gibson, Randall L., 117
Gilbert, James I., 115
Gilbert, Wallace H., 23
Gillmore, Quincy A., 51, 52, 54, 59, 62
Girard, L. J., 46
Godman, John M., 123
Gordon, George H., 62
Gorgas, Josiah, 143
Gould, Frank, 132
Gould, William J., 128
Gove, Jesse A., 23
Gracey, Samuel L., 72
Granger, Gordon, 124
Grant, Ulysses S., 16, 80, 100, 150, 158, 165
Gray, M. Martin, 35, 52, 53, 59, 61

Green, Johnny, 83
Greene, John, 24, 27
grenade, 74
Greyhound, 148, 149
Gross, Samuel David, 194
Gross, Samuel Weissell, 59, 61, 194
Grove, William Robert, 5
Grove battery, 18, 156, 159
guncotton, 171

Halleck, Henry W., 3, 17, 152, 188
Halsey, Charles F., 23
Hampton, Wade, 109, 110
Hardee, William J., 83, 87, 94
Hardman, Lyman, 92
Harper's Weekly, 54, 55
Harrison, Burton N., 111
Harrison, George P., Jr., 61
Hart, Henry W., 122, 123, 125, 126, 128
Hatch, John P., 138
Haupt, Herman, 133–37, 139, 141
Hawley, Joseph R., 61
Hays, Gilbert Adams, 162
Hazen, William B., 89, 90, 92, 94
Hearn, Chester G., 219n61
Hedley, Fenwick Y., 88
Heintzelman, Samuel P., 24, 26
Henry, Joseph, 5
Hibbert, Hugh Robert, 7, 8
Hickenlooper, Andrew, 84–88, 96, 97
Hight, John J, 107
Hill, Daniel Harvey, 22, 33, 37, 42–44
Hill, James H., 105
Hill, Moses, 24
Hindman, Thomas C., 146, 156
Hitchcock, Henry, 84, 85, 87, 88, 94
Hogane, James T., 45
Hogg, Harvey, 16
Holbrook, William C., 114
Hottle, James M., 140
Howard, Oliver O., 26, 27, 89, 93, 97, 109
Hubard, Robert E., Jr., 22, 31
Huntington, David, 92
Hyde, Thomas W., 30

Hyde, William L., 59

Illinois units: Battery I, Second Light Artillery, 96; Ninth Mounted Infantry, 142; Twelfth Cavalry, 23; Thirty-second Infantry, 88; Thirty-seventh Infantry, 120; Forty-eighth Infantry, 92; Ninety-fifth Infantry, 114; Ninety-seventh Infantry, 120, 121, 123; 116th Infantry, 92; 117th Infantry, 115

Illustrated London News, 159

Indiana units: Sixty-ninth Infantry, 122

Inhumane Weapons Convention, 183

Iowa units: Thirty-second Infantry, 115; Thirty-third Infantry, 115, 122

Jackson, Isaac, 124, 126

Jackson campaign, 46–49, 51, 169

Jacobi, Moritz Hermann, 6, 7

Jane Duffield, 152

J. E. Kendrick, 150, 152

Johnson, Nathan J., 104

Johnston, Joseph E., 22, 33, 38, 41, 43, 44, 46, 110, 169

Jones, Charles H., 109

Jones, David, 132

Jones, John B., 41, 42, 48, 62

Jones, Samuel, 82

The Journal of ERW and Mine Action, 184

Joyce, Fred, 47

Kellersberg, Julius, 65

King, W. R., xix, 11, 131–33, 170, 171

Knowlton, A. L., 101

Kochan, Michael P., xvi, 18, 102, 168, 193, 194

Korean War, 182

Lacey, Curtis P., 47

Lamb, William, 99–102

landmines, after the Civil War, 175–84; animals, 176; anti-vehicle mines, 177; booby traps, 177, 178; casualties, 180–82; civilians, 181, 182; electrical detonation, 176; moral revulsion, 177–79; number planted during Cold War, 183; number planted during World War II, 180; prisoners, 180–82; removal, 180–82; self-activated (pressure-activated, victim-activated, or contact), 176; tactics, 176, 179–81; technology, 177–81

landmines, before and during the Civil War: animals, 71, 78, 79, 110, 115, 117, 128, 194; booby traps, 14, 15, 24, 32–36, 116, 129, 140, 166, 189, 190, 204n31; bridge torpedoes, 133–37, 139, 171; casualties, xv, 29, 31, 47, 49, 54–56, 59, 61, 78, 92, 93, 116, 117, 125, 120–22, 128, 132, 150, 152, 193, 196; chemical detonation, 6–9, 127, 129; civilians, 33, 47–49, 71, 72, 132, 144, 150, 163, 173, 186, 190, 193, 207n16, 219n61; coal torpedoes, 147–49, 188, 217n62; covert operations, 131–53; definition, xvii–xviii; disturbance fuse, 4; doctrine, 32, 67, 70, 75–77, 185; electrical detonation, 5, 9, 10, 15–19, 100–103, 129, 142, 155–59, 171, 192; electrochemical detonation, 5, 6, 38, 155, 156, 159; electromagnetic detonation, 5, 6, 102, 103, 155–57, 159; emotional effect, 9–11, 32, 33, 76, 92, 94, 115, 120, 124, 125, 126, 143–45, 163–65, 168, 186, 190, 191; horological torpedoes, 148, 149, 152; injuries, 85, 88, 92, 96–98, 122, 127, 193–96; manual of landmine use (*see also* "Torpedo Book"), xiv; moral revulsion, xv, 10, 11, 20, 29–31, 34–36, 54, 55, 62, 64, 79, 85, 88, 89, 94, 108, 115, 117, 126, 147, 148, 161–63, 168–72, 186–90; number planted in Civil War, 76, 128, 129, 186; prisoners, 29, 30, 47–49, 54,

72, 85–88, 90, 93, 96, 107, 115, 118, 119, 123–26, 140, 144, 162, 178, 180–82, 186, 187, 189, 207n16, 209n37; railroad torpedoes, 137–39, 141–45, 215n15; removal, 28, 29, 47, 54–56, 72, 85–87, 90, 93, 94, 107, 108, 119 120 123–25, 140, 141, 186, 189, 191, 209n37; revenge, 34; Russia, xiii, 6–10; self-activated (pressure-activated, victim-activated, or contact), xiii, 1, 49, 155, 171, 186, 190, 192; sensitive primer, 41, 42, 46, 52, 53, 67, 74, 119, 127, 166, 168, 173, 185, 191; tactical use, 9, 48, 59, 61, 70, 75–77, 87–90, 92, 94, 100–102, 107, 108, 114–20, 125, 126, 129, 143, 144, 153, 163–66, 168, 190, 191; technology, 34, 48, 75–77, 79, 101–4, 107, 115, 116, 118, 119, 126, 127, 129, 133–38, 141, 146, 147, 153, 159, 162–65, 168, 190, 191; wood torpedoes, 145–47
Lathrop, D. B., 24, 27, 189, 200n13
Lay, George W., 40
Lee, Francis D., 15
Lee, Robert E., 38, 73, 142
Lewis, H. L. D., 116
Lewis, Joseph H., 83
Lewis, Josias, 123
Lexington, siege of, 13, 14
Liddell, St. John R., 116, 119, 212n14
Lieber, Francis, 188, 221n6
Lieber Code, 188
Livingston, Gary, 208n28
Lockett, Samuel H., 113, 117
Longstreet, James, 31, 32, 34, 79, 169
Lord, Francis A., xvii
Lovie, Henri, 17, 18
Lowden, Robert, 148, 149
Ludlow, William, 94, 96, 107

Mack, Thomas, 195
Magaw, Theophilus M., 90
Magruder, John B., 21

Maine units: Seventh Infantry, 27
Mallory, W. R., 124
Malta, 2
March to the Sea, 82–98
Marks, James J., 24
Marshall, Thomas B., 120, 125
Mason, A. P., 33
Mason, Edwin C., 27
Massachusetts units: First Cavalry, 70; Fifth Battery, 22, 24, 26; Twenty-second Infantry, 23, 24; Twenty-fourth Infantry, 54
Maury, Dabney Herndon, 68, 69, 113, 117, 167
Maury, Matthew Fontaine, 15, 38
Maxwell, John, 149–52
McClellan, George B., 21, 27, 28, 162, 186
McDaniel, Zere, 68, 69, 75, 141, 149, 152
McDermott, Michael, 23
McFarrar, George, 23
McMillan, Garnett, 109, 110
McNall, Thomas, 59
Mehaffy, Calvin D., 29
Meigs, Montgomery C., 135
Memphis Daily Appeal, 143
Merriam, Jonathan, 115
Mexican War, 2
Michie, Peter S., xvi, 56, 170
Miller, Monroe Joshua, 115, 126
Mills, John S., 122
Mills, Thomas J., 208n28
minefield, sapping through, 51, 53, 54
Minnich, John W., 35, 36
Minor, Carter Nelson Berkeley, 78
Missionary, 146, 149
Missouri units, US: Sixth Infantry, 92
Mobile, Alabama, xv, 42, 68, 69, 73, 74, 167
Mobile campaign, 113–29, 172, 173, 187, 192, 194, 212n14, 219n61
Moore, Elias, 122, 126
Moore, John, 92
Moore, Risdon, 115

Morgan, John S., 118
Mosby, John S., 73, 142, 143
Mower, Joseph A., 87
Murphy, W. R., 119

Napoleonic Wars, 2
Nashville Union, 30
New York Daily Tribune, 150
New York Herald, 29, 33
New York units: First Engineers, 138; Twelfth Cavalry, 110; Fortieth Infantry, 23; Forty-fourth Infantry, 23; 102nd Infantry, 107; 132nd Infantry, 132; 158th Infantry, 132
Nicholas I, 6, 10
Nichols, George Ward, 89
Nickels, Edgar A., 139
Nobel, Immanuel, 6, 7, 9, 127, 155, 175, 191
Nourse, Henry S., 48
Nowlin, Samuel H., 72

Øersted, Hans Christian, 5
Ohio units: Thirtieth Infantry, 92; Forty-seventh Infantry, 90; Eighty-third Infantry, 120, 124; Ninety-sixth Infantry, 116; 114th Infantry, 122
Oladowski, Hypolite, 147, 188
Oliver, John M., 209n37
Ord, Edward O. C., 79, 100
Osborn, Hartwell, 96
Osborn, Thomas W., 93
Osterhaus, Peter J., 143
Otto, John Henry, 86, 87

Pack, Reynell, 7–8
Paddock, Katherine Rains, 167
Palmer, Innis, 131, 132
Parker, James P., 46, 48, 49
Pennsylvania units: Third Heavy Artillery, 139; Fifty-second Infantry, 23; Sixty-second Infantry, 22; Sixty-third Infantry, 24, 162; Sixty-ninth Infantry, 24, 26, 28; Seventy-ninth Infantry, 107; Eighty-fifth Infantry, 23, 24
Perry, Milton F., xvi, 167
Perry, Oran, 126
Petersburg, siege of, 42, 74–80, 129, 158, 172, 196
petroleum torpedo, 160, 161
Phillips, Charles A., 26
Pleydell, J. C., 1
Poe, Orlando M., 89, 90
Polk, Leonidas, 15, 16
Pooler Station, Georgia, 83–89, 161, 162, 186, 207n16, 208n26
Port Hudson, siege of, 42, 45, 46, 63
Port Royal, South Carolina, 13–14
Porter, David D., 10–11, 146–48, 172
Prescott, George B., xviii
Prescott, R. B., 150, 152
Purviance, Harry A., 23, 24
Pyne, Henry R., 72

Rains, Gabriel J., xiv, 3–5, 19, 49, 67, 69, 70, 72–74, 82, 94, 102, 105, 106, 109–11, 113, 119, 129, 142, 147, 149, 152, 156, 179, 185, 187–89, 190, 192; administrator, 40, 41, 43, 69, 73; at Charleston, 53, 61, 62, 69; after Civil War, 155, 162–69, 171; family of, 39, 111; at Jackson, 46, 48; at Petersburg, 74–80; at Seven Pines, 37–39, 41; underwater mines, 38, 39, 42, 167; and Vicksburg, 43–46; at Yorktown, 21, 22, 27–28, 31–35, 37
Rains, George Washington, 39, 42, 81, 82
Randolph, George W., 34, 41
Rappahannock Station, Virginia, 215n15
Ratchford, James W., 22, 36
R. B. Hamilton, 167
Read, George W., 194
Rhode Island units: Battery B, First Light Artillery, 26; Third Heavy Artillery, 14; Fourteenth Heavy Artillery, 64

Rich, William H., 194
Riley, Edward J., 23
Riley, Peter, 195
Ripley, Edward Hastings, 79
The Rise and Fall of the Confederate Government, 166, 167
Rives, Alfred L., 69, 145
Roberts, Edward A. L., 160, 161
Robinson, Powhatan, 45
Roddey, Philip D., 142
Rodes, Robert E., 37
Roemer, Jacob, 47
Ross, Samuel, 96
Ruger, Thomas H., 165
Russo-Japanese War, 176, 187
Russo-Turkish War, 175
Rutherford, Kenneth R., xvi

Sanford, George B., 71
Santander, siege of, 2
Saunders, A. L., 16, 19
Savannah, siege of, 42
Scaife, H. F., 46, 47
Schafer, Louis S., xvii
Schaff, Morris, 150–52
Schiller, Herbert M., xiv, xvi, xix, 165, 185
Schirmer, William, 35
Schmitt, William A., 17
Schneck, William C., 128
Schubert, Justus, 53
Schweidnitz, siege of, 1
Scientific American, 18
Scofield, Hiram, 123
Scott, Henry Lee, 3
Scott, W. L., 118
Sebastopol, siege of, xiii, 6–10, 54, 76, 127, 155, 175, 185, 187
Second Seminole War, 4, 5
Seddon, James A., 44, 62, 69, 70, 143, 145
Seven Days campaign, 39
Seven Pines, battle of, 37, 38
Seven Years' War, 1
Seward, William Henry, 158

Shaffner, Taliaferro P., 157–60
Shannon, I. N., 146
Sheridan, Philip, 70, 72
Sheridan's Raid, 70–72, 78
Sherman, William T., 46–48, 81, 82, 85–87, 93, 94, 106, 139, 140, 143, 144, 161, 162, 186, 207n16
Simmons, William, 169
Singer, Edgar C., 63, 141, 145
Sister's Ferry, South Carolina, 106–8
Slocum, Henry W., 106, 107
Smith, Charles F., 137–39, 141
Smith, Henry I., 108
Sneden, Robert Knox, 22, 28
Snow, David Bassett, 120, 124, 129
Snyder, Dean, xvii
Sorrel, G. Moxley, 31–34
Southern Historical Society Papers, 166, 167
Southern Practitioner, 196
Spanish Fort, Alabama, 114–19, 126, 128, 173
Sperry, Andrew F., 117, 125, 173
Springfield Armory, Massachusetts, 217n62
Stanton, Edwin M., 158
Steedman, James B., 144
Steele, Frederick, 114, 119, 124, 186
Stinson, Jack Bryant, 78, 196
Stoneman, George, 135
Strong, William E., 89
Stuart, James E. B., 143
Submarine Warfare, 171
Sulakowsky, Valery, 65
Sullivan, John E., 69
Sultana, 148, 149
Susquehanna, 167
Suter, Charles R., 137–39
Sweitzer, Nelson B., 71

Tarleton, Robert, 69
telegraphy, 24, 189
Terry, Alfred H., 100, 101, 104
Terry, Carlisle, 147
Teterick, Harrison, 122

Tibbetts, Henry B., 132
Tielke, J. G., 2
Tomb, James H., 82–84, 87, 89, 90, 98, 107, 169, 208n26
torpedo, definition of, xvii–xix, 11, 175
"Torpedo Book," xvi, 33, 43, 46, 75, 76, 156, 162–67, 185
Torpedo Bureau, 39, 168
torpedo crimes, xix, 11, 31, 186, 187
Torpedoes: Their Invention and Use, 170, 171
Trowbridge, John T., 172
Tucker, John R., 83
Tullahoma campaign, 141
Tupper, Francis W., 84–88, 96, 97, 161

underwater mines, 9–11, 15–17, 38, 53, 57, 59, 68, 73, 74, 96, 131–33, 139, 167, 171–73, 177, 189
United States Military Academy, 10, 42, 157, 165
United States units: First Cavalry, 71; Third Colored Infantry, 54; Tenth Infantry, 26; Fifty-first Colored Infantry, 121
USS Baron DeKalb, 68
USS Black Hawk, 148
USS Cairo, 44, 68
USS Tecumseh, 73

Vance, Zebulon B., 45
Van Duzer, John C., 84, 85
Van Lew, Elizabeth, 78
Verdú y Verdú, Gregorio, 9
Vicksburg, siege of, 43–46
Vietnamese War, 182
Volta, Alessandro, 5
voltaic pile, 5

Wagner, George, 195
Wainwright, Charles S., 26
Walker, Richard H. L., 122
Walrath, Ezra L., 105
Walthall, W. T., 166

War of the Pacific, 175, 176
War of the Triple Alliance, 169
Warner, Thomas, 11
Washburn, Cadwallader C., 62–64
Webster, 148
Weeden, William B., 28
Westervelt, John H., 61
West Virginia units: Fourth Infantry, 145
Wheatstone, Charles, 6
Wheatstone electromagnetic battery, 103, 106, 159, 170
Wheeler, George, 65
Wheeler, Joseph, 143, 145, 146
White, Thomas S., 89, 93
Whitehead, Robert, 175
Whitesides, James, 122
Whiting, W. H. C., 42, 105
Wideman, John C., xvi, 18, 102, 168, 193, 194
Williamsburg Road, Virginia, 27, 28, 31, 33, 35, 41, 42, 48, 67, 163, 166, 169
Wilson, James H., 70
Wilson, John M., 115, 118, 127
Wilson, Joseph A., 14
Winder, John H., 140, 162
Wisconsin units: Thirty-fifth Infantry, 117
Wisewell, Moses N., 160
Wisner, James A., 120
Withers, William T., 68
Wollaston battery, 38
Wolters, Timothy S., xvii
Woods, Joseph T., 125
World War I, xiv, 175, 177, 187, 192
World War II, xiv, 178–81, 190, 191
Wright, Thomas J., 144

Yorktown, siege of, xv, 13, 15, 21–36, 37, 42–44, 49, 67, 76, 162, 169, 185, 187, 188, 190, 191, 194, 200n13
Youngblood, Norman, xvi, 171, 176

Zimmerman, Samuel, 1, 2
Zulu War, 176

About the Author

Earl J. Hess, professor emeritus at Lincoln Memorial University in Harrogate, Tennessee, is the author or editor of twenty-six books on Civil War history. They include *Civil War Logistics: A Study of Military Transportation* (2017), *Civil War Supply and Strategy: Feeding Men and Moving Armies* (2020), and *Animal Histories of the Civil War Era* (2022). Hess's study *Civil War Infantry Tactics: Training, Combat, and Small-Unit Effectiveness* (2015) received the Tom Watson Brown Book Award of the Society of Civil War Historians.